PLANNING OLYMPIC LEGACIES

When a city wins the right to hold the Olympics, one of the oft cited advantages to the region is the catalytic effect upon the urban and transport projects of the host cities. However, with unparalleled access to documents and records, Eva Kassens-Noor questions and challenges this fundamental assertion of host cities who claim to have used the Olympic Games as a way to move forward their urban agendas.

In fact, transport dreams to stage the "perfect Games" of the International Olympic Committee and the governments of the host cities have led to urban realities that significantly differ from the development path the city had set out to accomplish before winning the Olympic bid. Ultimately it is precisely the IOC's influence—and the city's foresight and sophistication (or lack thereof) in coping with it—that determines whether years after the Games there are legacies benefitting the former hosts.

The text is supported by revealing interviews from lead host city planners and key documents, which highlight striking discrepancies between media broadcasts and the internal communications between the IOC and host city governments. It focuses on the inside story of the urban and transport change process undergone by four cities (Barcelona, Atlanta, Sydney, and Athens) that staged the Olympics and forecasts London and Rio de Janeiro's urban trajectories. The final chapter advises cities on how to leverage the Olympic opportunity to advance their long-run urban strategic plans and interests while fulfilling the International Olympic Committee's fundamental requirements.

This is a uniquely positioned look at why Olympic cities have – or do not have – the transport and urban legacies they had wished for. The book will be of interest to planners, government agencies and those involved in organising future Games.

Eva Kassens-Noor is an assistant professor of urban and transport planning in the School of Planning, Design and Construction (SPDC) holding a joint appointment with the Global Urban Studies Program (GUSP) at Michigan State University. She is also an adjunct assistant professor in Geography. She received her Diplom-Ingenieur in Business Engineering from the Universität Karlsruhe (TH) in Germany, her MST (Master of Science in Transportation) from the Civil and Engineering Department at MIT, and her Ph.D. from the Department of Urban Studies and Planning at MIT.

PLANNING OLYMPIC LEGACIES

Transport dreams and urban realities

Eva Kassens-Noor

Routledge
Taylor & Francis Group

LONDON AND NEW YORK

First published 2012
by Routledge
2 Park Square, Milton Park, Abingdon, Oxon OX14 4RN

Simultaneously published in the USA and Canada
by Routledge
711 Third Avenue, New York, NY 10017

Routledge is an imprint of the Taylor & Francis Group, an informa business

British Library Cataloguing in Publication Data
A catalogue record for this book is available from the British Library

Library of Congress Cataloging in Publication Data
Planning Olympic legacies : transport dreams and urban realities / Eva Kassens.
    p. cm.
Includes bibliographical references and index.
1. City planning.   2. Municipal government.   3. Municipal services.
4. Olympics—Planning.   5. Olympics—Management.   I. Title.
HT166.K3595 2012
307.1'216—dc23

                                                        2011045342

ISBN: 978–0–415–68959–5 (hbk)
ISBN: 978–0–415–68971–7 (pbk)
ISBN: 978–0–203–11948–8 (ebk)

Typeset in Bembo
by RefineCatch Limited, Bungay, Suffolk

Printed and bound in Great Britain by
CPI Antony Rowe, Chippenham, Wiltshire

To my husband M. JehanZeb Noor,
our greatest blessing,
and our best friend

CONTENTS

TABLES

FIGURE

MAPS

ACRONYMS

ACOG	Atlanta Organizing Committee for the Olympic Games
AOC	Atlanta Olympic Committee
AOC	Australia Olympic Committee
ANC	Arees de Nova Centralitat = the New Center Areas
ARC	Atlanta Regional Commission
ASOIF	Association of Summer Olympic International Federations
ATHOC	Athens Olympic Organizing Committee
BIE	Bureau International des Expositions
BOA	British Olympic Association
BOC	Beijing Organizing Committee
BRT	Bus Rapid Transit
CBD	Central Business District
CETRAMSA	Center Metropolitan d'Informacio i Promocio del Transport SA
CIO	Comité International Olympique
CMB	Corporacion Metropolitana de Barcelona
CODA	Corporation for Olympic Development in Atlanta
COOB	Comite Organizador Olimpico de Barcelona '92
CTRL	Channel Tunnel Rail Link
DLR	Docklands Light Railway
EMT	Entitat Metropolitana del Transport
FGC	Ferrocarrils de la Generalitat de Catalunya
FIFA	Federation Internationale de Football Association
HOV	High Occupancy Vehicle
IBC	International Broadcast Center

IF	International Federation
IOC	International Olympic Committee
ISTEA	Intermodal Surface Transportation Efficiency Act
ITS	Intelligent Transportation Systems
IVHS	Intelligent Vehicle Highway System
LOCOG	London Organizing Committee of the Olympic Games
MARTA	Metropolitan Atlanta Rapid Transit Authority
MEPPW	Ministry of Environment Planning and Public Works
MPC	Media Press Center
NOC	National Olympic Committee
NSW	New South Wales
OAKA	Ολυμπιακό Αθλητικό Κέντρο Αθηνών (Athens Olympic Sports Complex)
OASA	Οργανισμός Αστικών Συγκοινωνιών Αθηνών (Athens Urban Transport Organization)
OCA	Olympic Coordination Authority
OCOG	Organizing Committee of the Olympic Games
ODA	Olympic Delivery Authority
OECD	Organization for Economic Co-operation and Development
OFFS	Olympic Family Fleet System
OFM	Olympic Family Members
OFTS	Olympic Family Transport System
OGGI	Olympic Games Global Impact
ORTA	Olympic Roads and Transit Authority
OSTS	Olympic Spectator Transportation System
OTS	Olympic Transport System
OTSG	Olympic Transportation Support Group
OV	Olympic Village
PERI	Plan Especial de Reforma Interior
PGM	Plan General de Metropolitano
PHF	Pan-Hellenic Federation
RTA	Roads and Traffic Authority
SOCOG	Sydney Organizing Committee for the Olympic Games
SRA	State Rail Authority
STA	State Transit Authority
TMB	Transports Metropolitans de Barcelona
TMC	Transport Management Center
TOK	Transfer of Knowledge
USOC	United States Olympic Committee
VIP	Very Important Person

FOREWORD

Legacy and its management are relatively new priorities of the International Olympic Committee (IOC). A commitment to legacy did not appear in the *IOC Charter* until 2003. Since then, it has become mandatory for a city to articulate its legacy vision and nominate specific plans about the management of post-Games legacy. Legacies, as the author suggests, are 'unsolved mysteries' in that there has been inadequate assessment and measurement of whether legacy objectives have been met afterwards.

The scholarly study of legacy has burgeoned in recent times with an increasing number of monographs and articles on this subject. These include general legacy works and legacy studies of particular Olympic cities. There have also been a number of works on particular themes such as cultural, economic, physical, psychological and social legacy. *Planning Olympic Legacies: Transport Dreams and Urban Realities* by Eva Kassens-Noor is the first major study of the evolution of transport legacies, from the 1992 Barcelona Olympic Games to the expected outcomes of the London 2012 and the Rio 2016 Olympic Games.

Transport has been one of the more problematic forms of legacy because of the competing demands of satisfying an Olympic peak of three weeks to long-term transport agendas integrated into a city's urban planning. While the IOC and Olympic organisers have tended to operate on a transport planning period of as little as three years, urban planners focus on transport developments over a period of decades.

The book documents the transport failure of Athens (1996) and shortcomings of Sydney (2000) and the lessons learned in terms of transport legacy. There is the hopeful development that the IOC is moving towards a broader and more sustainable approach to long-term transport legacies.

Eva Kassens-Noor has written an innovative and pioneering study that will enhance both legacy debates and the understanding of legacy. Her book will be of value to transport planners, Games organisers and scholars. *Planning for Olympic Legacies* has an impressive research base and is written in a clear and accessible style.

It is a privilege to write this Foreword and to commend this book to scholars involved in Olympic, transport and urban studies. Eva Kassens-Noor was an Australian Olympic Graduate Fellow at the Australian Centre for Olympic Studies in 2008 when she conducted research on Sydney's transport legacy. It is pleasing to see that her work has resulted in an authoritative chapter on this topic.

Professor Richard Cashman
Director, Australian Centre for Olympic Studies
University of Technology, Sydney

PREFACE

Large sporting events have fascinated me ever since I was a child, especially the World Cups and the Olympic Games. Born in a small village in southwestern Germany, my wishful dream of attending one such event came true when I was only 16: I received a young athlete award to visit the 1996 Atlanta Games. While I enjoyed the Olympic events in Atlanta, sprinting in particular, I also remember being stuck in buses for hours and taking the wrong bus after viewing a competition. Eventually, a volunteer had to escort me back to my athlete summer camp. I wondered about how much work must have gone into preparing for so many visitors in such a short period of time. That trip shaped my career – wanting to become a teacher or a sports-related professional, I chose a variation of both: to study sporting mega events, their related transportation requirements and their legacies as an academic.

This was just the beginning; indeed, my passion to follow sports and to understand mega events like the Olympics grew further over the next 15 years. As a Masters student, I had the chance of working for the International Airport in Athens during the European Cup finals and the 2004 Olympics. Guided by an international transportation expert and renowned professor, Amadeo Odoni, I co-led a small team of practitioners to set up the airport management plan for the Olympic family. Inspired by the positive impact, whereby the Athens Airport Authority was lauded for smooth operations, I decided to deepen my expertise with further targeted research and broaden my focus on metropolitan-wide urban and transport legacies.

The next few years took me to several Olympics cities – from the past and the future – as I strived to crack one key question: do the Olympics create a truly good, neutral or bad urban and transport legacy for host cities? The book draws from extensive field work in Barcelona, Atlanta, Sydney and Athens – and projects into the future for London and Rio de Janeiro. While I found very little academic work done with a singular and comprehensive approach to assess multiple host cities, I was fortunate in having ideal research advisors for developing my own holistic approach in doctorate work: Professors Karen Polenske and Chris Zegras from Massachusetts Institute of Technology, and Arnold Howitt from Harvard University. With their balance of regional economics, political economy, transportation and strategic planning perspectives, these advisors injected the rigor and insight in my research that it deserved. The next step then was to join the core Olympics research group at the University College in London – where I gained first-hand insight into London's Olympic dreams.

Another unique circumstance was my Fellowship with the International Olympics Committee (IOC) in Lausanne, during 2006. I had the unprecedented privilege of full access to the Olympics archives. Not only did I review meeting minutes and application files of multiple Olympic cities now covered in my work, but I also made valuable contacts at the IOC, which led to several high-profile interviews across Olympic cities. My interactions with Professor Philippe Bovy, IOC's lead transport advisor and global expert, turned out to be the single largest source of insight into the Olympic planning process. As I deepened my knowledge on behind-the-scenes forces that control the Olympics and define the resulting legacies, I also realized that this important field has only a handful of experts who advise the IOC and Olympic cities. This redoubled my personal drive to contribute towards positive Olympic legacies through actionable recommendations.

Over the last three years, I have further refined my findings and analyses at Michigan State University due to multiple opportunities crossing my path as an academic for transport during large sporting events. Based on my emerging profile and professional network, I was invited once again to arrange transportation plans for Olympic travelers: this time for the 2010 Vancouver Olympics. By far the youngest member of an internationally renowned expert team and the only academic, my recommendations were highly valued and implemented. A significant part of my work is now being developed and applied for other mega events. For example, I served on an expert team that advised the King of Saudi Arabia on transportation mass movement measures for the Hajj (Muslim pilgrimage in and around Makkah). These recommendations have largely been implemented. In related fields, I also suggested further applications for emergency evacuation planning in natural disasters and sustainable urban transport development. I still believe shaping Olympic legacies would be the single largest contribution I could make in my professional career to date.

What you will read in this book are untold secrets of applying and planning for the Olympics, the transport dreams with all the public hype, and the urban realities with broken or partially fulfilled promises. There is no one entity or single person to hold accountable. In addition, the IOC and future host cities are now learning how to adopt their approaches to have positive transport legacies. You will see that host cities pay for the Games through their sweat, tears and coffers for generations to come; however, the question of what they win in return has been largely unsolved. I am not claiming this book is the holy grail. It is only an attempt to demystify the legacy and impact of the Olympics, especially from the urban transport and metropolitan development angle. My audience is any interested planner, academician or even resident who wants to shape the future legacies by leveraging simple lessons learned and key principles – without having to fully or blindly believe what their national or IOC officials claim as a must-have. My aspiration is for both the IOC and future Olympic planners to be well aware of how to avoid the mistakes of the past to build a better future together.

I hope you will enjoy, as much as I have, this privileged Olympic journey, starting from the application stages to planning processes, eventually all the way to revisiting legacies. You will discover first-hand insights from the past and future host cities, where I lived, conducted deep structured research and inter-viewed dozens of practitioners who have devoted their life to make Olympic legacies a success – some of them working with me from well into their own retirements. This work would not be possible at all without their generosity and mentorship.

And as a young, still ever-so-curious researcher and academic, my journey is just beginning. I can only hope to contribute towards bringing my predecessors' transport dreams a few steps closer to urban realities – with your support and interest, either as a planner or reader, along the way.

Eva Kassens-Noor
East Lansing, Michigan
September 2011

ACKNOWLEDGEMENTS

I simply could not wish for a better group of colleagues, planners and individuals who contributed so selflessly to this book. Thank you!

I would like to especially thank Philippe Bovy, Professor at EPFL and IOC transport advisor, without whom this work would not have been possible. Because he deeply cares about the Olympics bringing a positive legacy for its host cities, he offered me access to his personal documents, strongly supported my research fellowship in Lausanne, and established first contacts to prestigious and key Olympic personnel across cities and institutions. I am also deeply indebted to Professors Karen Polenske, Arnold Howitt and Chris Zegras, who rigorously helped me think through my research.

My research conducted across six countries would not have been possible without the support of academics, scholars and researchers at the local Olympic centers, planning archives and universities. In Lausanne, I especially owe thanks to Nuria Puig, who found ways to waive the 25-year research and publication embargo on the Olympic host cities for me. In Barcelona, I thank Merce Rosello and Berta Cerezuela for providing mè access to the Olympic archives at the University of Barcelona. Furthermore, I would like to thank Lluís Millet for his advice on my Barcelona work. In Atlanta, my gratitude belongs to Sue Verhoef for pulling the necessary documents from the special Olympic collections at the Kenan Research Center. Furthremore, I would like to thank Professor Greg Andranovich for his encouragement to pursue this book project. In Sydney, special thanks belong to Professors Richard Cashman, John Black and Graham Currie. All individuals have significantly contributed to my understanding of transportation legacies, supported me throughout my Olympic journey, and helped me gain access to the local archives focused on the Olympic Games. In Athens, I thank Dr. John Frantzeskakis for all his help, support and providing me with multiple interviewee contacts to gain a holistic understanding of the planning processes. In London, my gratitude belongs to Professor Harry Dimitriou, who welcomed me into his research group and provided guidance for my book. In Rio de Janeiro, I thank Professor Chris Gaffney, who shared valuable insights about the ongoing planning processes right up until this manuscript was submitted.

Above all, I owe thanks to all my interviewees for their precious time during my interviews and for providing me with valuable information about the Olympic planning processes and resulting legacies. These urban planning and transport experts provided me with insights on the transport planning process for the Olympics despite their personal obligations. Furthermore, I would like to thank my personal editor and writing consultant, Bob Irwin, as well as the members of my writing group, Professors Trixie Smith, Janice Molloy, Manuel Colunga and Wen Lu.

My deepest gratitude belongs to my husband, JehanZeb Noor. He has patiently supported me throughout my Olympic journey and made my life brighter every day.

1

THE OLYMPIC GAMES

An ephemeral opportunity for cities

Olympic legacies are one of the unsolved mysteries surrounding the aftermath of the world's largest sporting event. While many host cities plan for valuable legacies, what exactly they are and how they came into being, are widely debated questions among international scholars, politicians, urban planners and sports enthusiasts. True legacies only emerge after the Games and remain largely unexplored as the spotlight of the event shifts to the next host. One fact, however, is indisputable: the Olympic Games provide the host city with an unprecedented opportunity for change. Within a few years, cities undergo a transformation stimulated by the hope for economic growth, a world-class image, enhanced connectivity and urban regeneration. Yet whether these desirable outcomes or their negative side-effects remain in host cities after the Games have left are "known unknowns"* (Horne 2007: 86). In this book, I appraise the Olympic Games and their effects on metropolises as momentous opportunities to realize ambitious urban and regional goals. I find that the International Olympic Committee (IOC) can influence these local goals, sometimes creating – but also squandering – valuable opportunities for the host city. Unless local governments take proactive steps to ensure that the urban and transportation decision-making does not serve only the event's short-term needs, city planners will not get the transformation and long-term benefits they had hoped for.

Yet it is those hoped-for legacies that are the typical rationale for most residents to back the bid. In a nutshell, legacies are consequences of mega events. Pioneering works on legacies categorize outcomes of the world's largest events (see Cashman and Hughes 1999; Clark 2008; Gratton and Preuss 2008; Poynter

---

* Known unknowns are impacts we know are going to be there after the event, but we do not know exactly what they are or will be.

2004; Preuss 2004; Ritchie 1984). The most common categories are cultural (Roche 2006), economic (Preuss 2004), political (Burbank et al. 2001), physical (Essex and Chalkley 2008), psychological (Malfas et al. 2004), and social (Horne and Manzenreiter 2006). Olympic legacies penetrate different geographical spheres, and can be positive or negative, beneficial or detrimental, tangible or intangible, visible or invisible. Their basic characteristics however, are intrinsically the same: they remain in the host cities' Umwelt for several subsequent years, decades and sometimes even centuries. Among these legacies, the theme of urban renewal has gained exploding importance as more and more local governments attempt to use the Games as catalysts for urban regeneration. For this book, I focus on the process through which legacies were created in transport and urban systems paying particular attention to those influences that are broadly stimulated by the Olympic Games as an event; these influences were either imposed by the IOC via explicit and implicit requirements during the pre-Olympic preparation stages, or – equally important – by city governments driven by ambitious aspirations to win and stage the "best Games ever."*

By examining the urban planning trajectory of Barcelona, Atlanta, Sydney, Athens, London and Rio de Janeiro with a particular lens on the IOC, I hope to provide a new perspective and a better understanding of the urban and transport legacies remaining in cities post-Olympics. The book draws on a unique set of data, given that I had access as an IOC Fellow to the archives containing the internal communications between the IOC and the hosts. These archives are under a 25-year embargo. With time and persistence, I also established a working relationship with the only transport advisor, appointed by the IOC in 1999, tasked to evaluate all Olympic bids in terms of transportation. With his assistance, I was able to conduct personal interviews with planners in charge of transport and urban development in Olympic host cities. Continuing my Olympic journey, I lived in Barcelona, Atlanta, Sydney, Athens and London for several months each, conducting interviews, archival analysis and site visits. I cross-compared and traced the creation of urban and transport legacies over several decades before and after the Games. In essence, I believe there are two important stories to tell.

The first is about the potential power that the IOC can exercise on host cities, thereby influencing the creation of legacies. The typical rhetoric surrounding the decision to bid for the Olympics almost always emphasizes the legacy impacts of the Games, but the dynamic that I document in this book is that cities frequently give priority to satisfying the short-term demands. Location choices for Olympic activities coupled with transportation demand peaks require certain alterations to metropolitan regions. Therefore, the IOC sets forth requirements to which

---

* The "best Games ever" is the usual phrase with which IOC presidents have ended the Olympic Games during the closing ceremony, in particular A. Samaranch. With the transition to J. Rogge, the phrase started to fade away.

bidding cities adhere, but frequently exceed. Either way, infrastructural invest-ments have become a necessity to stage successful Games. The degree to which Olympic requirements make the cities deviate from their original master plan, however, varies greatly. I attribute this degree of deviation to three factors. The first factor is that over time the IOC became more knowledgeable about handling Olympic passenger travel because of previous hosts' mistakes and successes. Hence, the committee imposed additional and more stringent requirements on cities over the past three decades. The second factor is that the more the IOC doubted the city could handle the Olympic transport, the more pressure it exer-cised on the city to comply with their requirements. Through this process the host cities created Olympically driven legacies. Even though the introduction of the IOC's requirement for bidding cities to justify expected legacies beyond the Games could have reversed the trend of Olympic legacies that do not fulfill long-term functions beneficial to the host city, Rio de Janeiro's bid lends evidence to the contrary. This leads me to the third factor: some host cities have ignored much of their strategic urban planning goals in favor of winning the bid. This might be due to not having an urban master plan in place or focusing on other legacies, such as economic growth or a global image, at the expense of urban legacies.

The second story is about the careful integration of Olympic requirements with urban master plans to create long-lasting and beneficial urban and transport legacies. Past host cities frequently failed to take long-term advantage of the Olympic event, notwithstanding what they originally thought they would do. To the extent that cities have been able to gain sustainable advantage from hosting the Olympics, it is because there has been an alignment of their long-term plan-ning perspective and the way they used the Olympics to improve their transport systems. My measure of alignment (between the IOC requirements and the metropolitan strategy) is the extent to which the city deviates from its original plans because of the Games.* Any deviation from the original city plans provides evidence for the role the Olympics play in creating legacies. The deviation can be driven by both: the host cities' new Olympic priorities and the IOC require-ments. I acknowledge that there might be other influences that change the orig-inal city plans in the run-up to the Olympics. This possibility, however, is highly unlikely due to the host city contract† and the immovable deadline the city faces in preparing for an event of such magnitude in only seven years. My conclusion is that if cities are not very careful to create and protect their goals, then the Olympic urban legacy will actually be quite minimal.

---

* Master plans developed before the city was awarded the right to stage the Games.
† The host city contract is an agreement signed with the IOC by the city hosting the Olympics, in which the city agrees to implement all modifications in the manner stated in the bid file. Changes to this agreement are difficult to implement and only possible with the consent of the IOC.

The creation of urban legacies

Without a doubt, the Olympic Games affect urban planning, while the legacies created in the process of staging the world's largest mega event remain in their host cities for decades. Particularly influential for residents are the infrastructural legacies in the host's urban environment as the city gets permanently trans-formed. Especially, "the Olympics . . . have an immense impact on the urban image, form, and networks of the host" (Short 2004: 25). Scholars (Coaffee 2007; Gold and Gold 2007; Jago et al. 2010) along with the IOC, frequently reiterate the importance of planning infrastructure not only for the mega event itself, but also for the city's urban trajectory to avoid white elephants.* The message for planners is clear: use the Games' catalytic momentum for urban development irrespective of the Olympics.

The desire of host cities to create urban legacies for spatially focused regen-eration and upgraded transportation networks is not a new phenomenon. Over the last half century, cities have endeavored to recreate their urban spaces using the Olympics as a powerful catalyst for urban change. Historically, the first Olympic Games to have a metropolitan-wide impact were the 1960 Rome Olympics. The bi-polarity of the two Olympic Zones (Foro Italico and Zona E.U.R.) stimulated architectural creativity and required coordination across city quarters. The following two Games, Tokyo '64 and Mexico City '68, were staged in mega cities and by far exceeded the logistical demands of previous Games on the hosts. Munich '72 and Montreal '76 were conceptually similar; both wanted to shine with new urban concepts and new technological innova-tions. And both encapsulated the Olympics away from their metropolitan region. Montreal's investments in massive Olympic constructions resulted in a billion-dollar deficit to the city, thereby having a deep impact on the following two hosts. Moscow '80 and Los Angeles '84 basically remained on their urban trajec-tory, allowing the Olympics little interference in their future urban develop-ment. Reviving the power of the Olympics in restructuring urban areas, Seoul '88 rediscovered the true potential of the catalytic Olympic effect, exceeding any previous urban development efforts in scale and magnitude.

As pioneers in their analysis of the Olympic Games and urban development, Stephen Chalkley and Brian Essex distinguished four phases of urban infrastructural development related to mega events over the past 100 years, showing an increasing scale of impact, organizational effort and international attention (Chalkley and Essex 1999; Essex and Chalkley 2004). At the same time, they also observed an

---

* White elephants are considered valuable possessions that their owners cannot discard. Their maintenance cost exceeds by far their value. The phrase originated in South East Asia (Thailand, Burma). In the 1800s, receiving a white elephant as a gift was both a blessing and a curse: a blessing because the white elephant was sacred, and hence had to be kept; a curse because the animal had to be worshiped, fed and could not be put to practical use to offset the cost of maintaining it.

increasing mismatch between the promises cities are willing to make in the pursuit of hosting a mega event and what the city needs for its future. The cities, given the imminent publicity and world-wide interest focused on the Games, "invest extravagantly in unnecessary infrastructure" (Essex and Chalkley 2002: 12).

Following these early explorations, attention has shifted away from infra-structure (except stadiums) towards the political process among local govern-ments and grassroot groups to seize the event opportunity for their own good (Cochrane et al. 1996; Owen 2002). For example, Preuss (2004: 1) noted that mega events offer a "unique opportunity for politicians and industry to move hidden agendas" by fast-tracking the decision-making process (see also Cashman 2002; Hiller 2006). This frequently leads to the exclusion of public participation (Hall 1989; Lenskyj 2002). Along with this exclusion came the shift of public resources from long-term investments like education to short-term needs of the event, such as imaging functions (Hall 1996; Jago et al. 2010; Searle 2002). A further controversy in using mega events as a strategy for urban renewal revolves around the discussion of the beneficiaries. Academics have argued that rarely have any improvements benefited citizens (Andranovich et al. 2001), and if they did, they benefited the wealthy (Eitzen 1996, 2002). Hence, Ruthheiser (2000) and Lenskyj (2002) suggested that mega events as such may serve to exacerbate social problems and deepen existing divides among residents.

Focusing back on urban development legacies, John R. Gold and Margaret M. Gold's (2007, 2010) ground-breaking books revive the discussion on urban development, city agendas and legacy creation. By retelling the stories of Olympic host cities, some researchers reflect on the regenerative power that the Games can have on urban areas (for example, see Coaffee 2007; regeneration also empha-sized in Poynter and MacRury 2009). The host cities' desire to use the Games to revamp tangible legacies, in particular urban regenerations, has become a main-stream topic in the legacy debate. So far the literature has ignored a detailed view on the role of the IOC in planning these desired legacies, because access to data, interviews and non-IOC-controlled insights are incredibly hard to come by. Frequently described as exercising "shadowy geopolitics" (Cochrane et al. 1996: 1323), or as a "self-perpetuating oligarchy immune from democratic account-ability" (Short 2008: 325), the intriguing processes of why and how the IOC can influence city planning has largely remained unexplored.

The creation of transport legacies

In particular, transport requires significant planning efforts, and can – if well planned – remain as a lasting legacy in hosting cities (Hiller 2006). The Olympic Games, like other extreme events, set passenger peak demands on transport systems of host cities (Kassens-Noor 2010a). Additional capacity is needed for millions of people, as Olympic peak loads can reach up to 2 million additional person trips per day (Bovy 2009). To manage these passenger peak demands, new transport agencies and communication structures among local transport agencies are created and new operational policies implemented, using a variety of traffic and transport management strategies (Kassens-Noor 2010b). During the Olympic Games, a commuting shift to public transport requires a myriad of additional cars, trains and buses (Hensher and Brewer 2002; Minis et al. 2009). In addition, the transportation system becomes highly vulnerable and contingency plans are a prerequisite for staging the event successfully (Minis and Tsamboulas 2008). Despite hopes to alter residents' travel behavior in the host cities, Olympic short-term changes usually do not translate to permanent ones. While strides have been made to uncover the operational complexity of Olympic transport, research on accessibility and mobility for special events is still in its infancy (Robbins et al. 2007).

Infrastructural investments in transport systems to prepare for Olympic peak demands have become a necessity to successfully stage the modern Games. Mega events lead to investments in public transportation, road construction and airport expansion; in return these necessary improvements provide the host city with a better quality of life, significant travel time savings, more transport capacity and more attractive transit modes (Bennett 1999; Christy et al. 1996; Getz 1999; Preuss 2004; Syme et al. 1989). Therefore, events can be planning instruments for urban transport development, such as clearing congested areas or re-organizing transport systems (Rubalcaba-Bermejo and Cuadrado-Roura 1995). Improved transport is a potential benefit frequently mentioned in bidding files and used as a claim to counter potential costs and burdens to the hosting community (Cashman 2002). In sum, transport systems need to be prepared for the Games, whereas the necessary alterations of the transport network remain as tangible legacies in host cities. These legacies can potentially enhance urban transport to accommodate future growth and be a stimulus to change the way people travel.

Planning Olympic legacies in four hosts and two future hosts

To date, the Games not fitting a planning model (Essex and Chalkley 2005), the frequent creation of white elephants (Gold and Gold 2007) and the involvement of the IOC in creating host legacies (IOC 2007) merit further inquiry. Accordingly, this book sets out to answer three key questions: (1) Is the urban planning process altered due to Olympic requirements? And if so, how? (2) Consequently, what role does the IOC play in urban development and the creation of transport legacies? (3) Given the IOC's role, how can host cities plan for valuable Olympic legacies and aspire not only to host the best Games ever, but also retain the best legacies ever? To take advantage of the opportunities mega events bring to cities, the key question for planners to answer is how cities can align the necessary transport provisions for staging a world-class event with their metropolitan vision, and hence, use the mega event as a driver towards desirable change.

In the following chapters, I investigate the urban planning processes and the resulting legacies of four Summer Olympic host cities, Barcelona (1992), Atlanta

TABLE 1.1 Olympic host cities' characteristics

Characteristic	Barcelona	Atlanta	Sydney	Athens	London	Rio
Country	Spain	USA	Australia	Greece	UK	Brazil
Year of the Olympics	1992	1996	2000	2004	2012	2016
Year of election	1986	1990	1993	1997	2005	2009
Year planning began	~1979	~1986	~1990	~1988	~1995	~1996
Tickets sold (in millions)	3	8.4	6.7	3.6	~7.7 (plan)	~7 (plan)
Number of athletes	9,400	10,400	10,600	10,600	10,600 (plan)	10,600 (plan)
Olympic main areas	4	1	1	3	1	4

Sources: Compiled by author. Adapted from ACOG 1994; ATHOC 2004; 2005, Bovy 2005, 2011; CMB 1986; Currie 2007; Greater London Authority 2011; House of Commons 2003; LOCOG 2005; OASA 2009; ODA 2009; ORTA 2001; Rio 2016 2009.

(1996), Sydney (2000) and Athens (2004), and foreshadow the planned legacies of two future host cities, London (2012) and Rio de Janeiro (2016). Even though all case cities are intrinsically different with unique histories, economies, political institutions, urban forms and transport networks (Table 1.1), they approached the Olympic planning process with the same goal: to stage successful Games.

While the number of athletes stayed fairly constant overtime, the tickets sold (or expected to be sold) varies greatly. Planning for the Olympic Games also had different time horizons and resulted in the creation of one central or multiple Olympic main areas. Therefore, each city's leaders, trying to use the Olympics as an opportunity for urban change, considered individual transport and urban goals in the city's spatial context. While my case study analysis stretches roughly over two decades, focusing on the three key questions about requirements, the IOC's role and legacies, will succinctly guide the internal structure of each chapter and the book.

Book structure

In Chapter Two, I introduce the International Olympic Committee (IOC) as a powerful stakeholder in the urban planning process. Throughout this chapter, I identify requirements the Olympics bring to bear on the transportation and venue planning process. My key argument is that the IOC, equipped with the Olympics, has the vision, the power and the necessary tools to intervene in the urban planning process.

The following four host city chapters (Chapters Three to Six) analyze the process for planning and implementing Olympic transport and urban legacies. The goal of each chapter is to highlight the role the IOC and host cities play in the planning process for urban and transport change. Each city chapter consists of four parts that together trace the city's development in chronological order. The first lays out the master plan for urban and transport development irrespective of the Olympics. The second analyzes the Olympic bidding and candidature plans. The third provides a quick glimpse into the time during the Games. The fourth reviews the legacies that manifested themselves in the urban realm and retrospectively analyzes missed opportunities as identified by Olympic planners.

In Chapter Seven, I review the planning processes of two future host cities, London and Rio de Janeiro. First, I provide an inside view into the master plans prior to the bid. Second, I trace the candidature proposals and their expected legacies. Third, I provide a glimpse into the projected legacies for both hosts in reflecting on the IOC's involvement.

In Chapter Eight, I define the role the Olympic Games have played, currently play and will likely play in urban development. For previous hosts, the exclusive focus on the Olympic Games has led to hasty decisions. This situation might change due to the new IOC questions about expected legacies; pre-planning for post-event usage has become an important evaluation criterion for the bid. Finally, I provide recommendations for potential hosts to achieve transport and urban dreams in the long term.

2

THE IOC AS A POWERFUL
STAKEHOLDER IN THE
PLANNING PROCESS

The International Olympic Committee (IOC) can have a significant effect on urban and transport developments of host cities. Essentially, the IOC is an agent of urban change, because it has an Olympic vision about how the city should function during the Games and possesses the institutional power to implement necessary changes to achieve such vision. City governments play the significant counterpart in realizing Olympic ambitions. As bidders aspire to win, they might introduce new urban goals and promise legacies that seem to be aligned with Olympic requirements yet never entered the planning process till the dawn of the Olympic candidature. Regardless of either influence in shaping actual legacies, the Olympics and other mega events hold significant power in restructuring cities. For example, Harvey (1989) suggested that urban spectacles have become a key element of urban and economic policy. A decade after his observation, Chalkley and Essex (1999) found similar tendencies and argued that mega events have turned from a tool to an agent that plays a significant role in urban policy. In line with their suggestions, Short (2004: 107) believes that "the Games act as an important tool to literally reshape the city, in both discursive and spatial terms."

This chapter introduces the IOC as a stakeholder in the planning process and attempts to argue the case for two important Olympic planning concepts. First, the IOC has the institutional power and necessary tools to influence the host cities' urban planning process. Second, the IOC's vision about the Olympics essentially implies a vision for the city. In the end, one story shapes this chapter. Throughout the bidding, candidacy and preparation stages, the IOC essentially superimposes an organizational structure for transport on cities via explicit and implicit requirements. Intrinsically, cities view these requirements as "golden rules" which, when complied to, enhance their chances in winning the right to stage the Games. Even though the requirements imply costly investments in transport and urban structures, legacy expectations are the hook for city residents to support the Olympic idea.

Having a vision and the power to implement it, would make the IOC a global actor of change with the ability to affect local urban development. Therefore, the first part of this chapter crystallizes "the vision" the Olympics bring to city development. The IOC's primary focus lies on an Olympic transport ideal of smooth operations for all Olympic client groups during the event. As a secondary aspiration, the IOC hopes that Olympic legacies benefit the cities' long-term development. The second part of the chapter puts forth evidence that the IOC has the power, ability and necessary tools to intervene in city-specific urban and transport planning processes to realize its Olympic transport vision. For urban structures, potential influences include the IOC preferred layout in nucleating Olympic activities. For transport, potential influences include accessibility requirements for stadia, mobility requirements for Olympic client groups and further transport planning recommendations. Before launching into a detailed analysis of the IOC as an agent of urban change, a brief introduction to the IOC, the Olympic Games and its transport challenge is necessary.

Introducing the Olympic Games and the IOC

The Olympic Games have become the largest sporting event in modern times. Baron Pierre de Coubertin re-inaugurated the Olympic Games in 1896 according to their Athenian ancient predecessor. Two years prior, Coubertin had founded the International Olympic Committee (IOC) as the governing body for this worldwide athletic competition. Since then, the modern Olympic Games take place every four years. The IOC itself defines its role in the organization of the Olympic Games as follows:

> The organisation of the Games consists of a partnership between the IOC and the Organising Committee for the Olympic Games (OCOG). The Games are the exclusive property of the IOC, which has the last word on any question related to them. The IOC plays a supervisory and support role; in other words, it controls the organisation of the Games, ensures they run successfully, and checks that the principles and rules of the Olympic Charter are observed.
>
> *(Olympic Movement 2008)*

The IOC awards the Olympic Games to a city after a careful process of evaluation (IOC 2003: Rule 36, para. 3). Like many world mega events, the Olympic Games are attributed by the event owners (IOC) to the event organizer (city) after a selective bidding process. For the event organizer, bidding entails defining objectives, structure and content of the event including its most significant requirements and constraints. For the event owner, bidding is an evaluation and decision process. The main aim of the bidding process is to facilitate selection of the most efficient and reliable project (Bovy 2005). The detailed procedure leading to the election of a host city for the Olympic Games is manifested in the Olympic Charter (IOC 2011: Chapter 5, Rule 33 and its bye-laws). This procedure is briefly sketched out below.

Olympic bidding is a two-stage, two-year process, preceding the actual Games preparation period of seven years. The first stage is called the bidding stage, in which prequalification for potential acceptance is evaluated by the IOC. One Olympic host city proposal per country is allowed, presented by the country's National Olympic Committee (NOC). If more than one city within one country aspires to bid, national elections are held and evaluated by the respective NOC. The second stage is called the candidacy stage. The candidacy dossier follows 17 criteria (called themes) set forth by the IOC (as of 2011): 1. Olympic Games concept and legacy; 2. Political and economic climate; 3. Legal aspects; 4. Customs and immigration formalities; 5. Environment and meteorology; 6. Finance; 7. Marketing; 8. Sport and venues; 9. Paralympic Games; 10. Olympic Village; 11. Medical and health services; 12. Security; 13. Accommodation; 14. Transport; 15. Technology; 16. Media operations; and 17. Olympism and culture.

The IOC carefully evaluates individual city proposals in comparison with other bids according to the Olympic Charter (IOC 2007: Rule 34). Only this full candidature can lead to the selection as a host city. The candidacy file turns into a host city contract between the IOC and the city upon awarding the city the right to host the Olympic Games. Within the following seven years preceding the Games, these candidacy plans shall not be altered, or, if they are, modifications require the consent of the IOC. The preparation of this host city contract is a multimillion dollar project.

The Olympic transport challenge

Efficiency, effectiveness and safety are the three keywords used in planning for mega events like the Olympics. Mega events are regarded as an outstanding transport challenge for the city and require broad strategic and operational measures (Bovy 2006; ECMT 2003). Recommendations include using the regional infrastructure to its maximum capacity and serving the visitors mostly through public transport services. Bovy and Liaudat (2003) provide a description of the mega event transport task. In their book, they distinguish six client groups (spectators, athletes, logistical staff, volunteers, VIPs, media), each of which has specific mobility and accessibility needs. Among client groups an implicit hierarchy is established. In order for an event to be successful, each of these clients' needs has to be fully met.

The function that Olympic transport assumes is to connect competition and training venues (sports stadia) with non-competition venues, such as the Olympic Village, hotels, the airport, and so on (Bovy 2005). The IOC distinguishes competition, training and non-competition venues. Competition venues are stadia in which Olympic sports take place, training venues are practice stadia, and non-competition venues are central points of interest within the Olympic system, such as the airport, the media center and the Olympic Village. The Olympic Village accommodates the athletes during the time of the Olympic Games. Munoz (1997) argued that this village should fulfill a dual function. Besides accommodating athletes, its design and structure should be considered for after-event usage. Up until the 1970s, the villages had been transformed into social or student accommodations. In recent years, they have been used as high-quality real estate. Munoz (1997) observed that this change corresponds to the parallel evolution of athletes: from a Spartan lifestyle to a tourist requesting the best amenities.

The Olympic transport task is an immense undertaking (as of 2010): 28 sports, more than 300 competition events, 16,000 athletes and team officials from 202 countries, over 5,000 IOC, NOC and International Federation (IF) members, over 20,000–30,000 media, over 30,000 sponsors and guests, over 50,000 volunteers, between 120,000 and 150,000 workforce staff members, between 4 and 9 million ticketed spectators, and between 2 and 13 million residents to be transported to multiple destinations within host city's region during three short weeks (Ostertag 2010). Combined, these travelers create extraordinary traffic peaks frequently reaching 1.25 to 2 million Olympic person trips per day (Bovy 2009). Transport's role, as the crucial backbone of the city, is to maintain stable linkages and ensure that almost all other major Olympic city functions can operate unimpeded (Bovy and Protopsaltis 2003). At the same time, Olympic transport has to maintain normal traffic conditions for residents. Transport has to fulfill six fundamental objectives according to the IOC transport consultant Professor Phillipe Bovy (2004a: 19–20):

- Transport must be safe and secure.
- Transport must be "absolutely" reliable.
- Transport must be efficient, comfortable, convivial and must guarantee short travel time especially for athletes and the media.
- Transport must be flexible to mitigate risk of interruptions.
- Transport shall be environmentally friendly.
- Transport shall contribute to a strong host city and regional mobility legacy.

Transport systems during the time of the Olympics must perform extraordinarily well. For Professor Bovy and the head of the Olympic transport management team in Athens, Panos Protopsaltis, mis-execution of the transport task carries high risks, because dysfunction can result in vivid criticism by the international media (Bovy and Protopsaltis 2003). Failure to perform perfectly can lead to severe congestion, overcrowding and athletes missing competitions. These outcomes can severely damage a city's global image (Rushin 1996).

Olympic transport requirements as a vision for the host city

Cities theoretically have power over two issues in implementing measures to comply with the IOC transport priorities: the land-use choice for venues and the provision of transport services for Olympic client groups. Their power, however, is heavily constrained by a superimposed IOC transport structure and "known" best practice solutions. The IOC's vision about the Olympics implicitly imposes a vision for the city in regards to transportation. The IOC's primary focus lies on the perfect functioning of the city's transport system: ensuring smooth operations for all Olympic client groups during the event. Therefore, cities must adopt certain transport priorities when hosting the Olympics (Bovy 2004a: 47–9): athletes, technical officials, broadcasters, selected VIPs (Priority 1); media, press officials, judges and specific workforce, which has to arrive well before the competition (Priority 2); National Olympic Committees (NOC), International Federations (IF) officials and Olympic Family members, who are directly associated with the athletes including family members, team coaches, personal trainers and so on (Priority 3); high-ranked Games sponsors, such as Coca Cola, McDonalds (Priority 4); workforce, volunteers and spectators (Priority 5); the general public (Priority 6). To ensure the level of transport services according to this hierarchy, the IOC sets forth mobility and accessibility requirements on hosting cities. To win the bid, cities have to compete with their transport concepts among other Olympic objectives. Because the bid turns into the host-city contract, cities are locked into their promised transport performance and have to deliver at Games' time. This way, the IOC's Olympic vision for transport becomes reality.

Mobility in the city and accessibility to venues

Venues are the central entry and exit points of mega events. The accessibility requirements for athletes and visitors alike to these venues are crucial. Compulsory by IOC requirements, each of the venues has to have two distinct functional areas for athletes and spectators strictly separated from each other. Both areas also must have two clearly distinct access routes due to security; the "front of house" (FOH) traffic and the "back of house" (BOH) traffic (Bovy 2004a: 52). The FOH traffic includes ticketed spectators and sponsors with no access to individual vehicles. The BOH traffic consists of transport priorities types 1 through 4 and fulfills emergency functions (Bovy 2004b). Around each venue, a security perimeter separates "city domain" from "mega event domain" (Bovy 2004b).

Connectivity of venues is a further crucial element in planning transport for the Games. The bidding questionnaire requires the cities to report travel distances between competition and non-competition venues (IOC 2000: 46). An estimated travel time matrix has to consider congestion levels, proximity to public and exclusive transport, and future infrastructure developments (Bovy 2004a: 33). Answers to the transport questionnaire also include a preliminary budget and the institutional structure of the Olympic Organizing Committee for the Olympic Games (OCOG). The bidding file further needs to lay out the relationships among public authorities in charge of transport, security and the environment (Bovy 2005).

The travel time matrix and the distance matrix are crucial evaluation criteria, by which cities are rated (IOC 2000: 46–7). Furthermore, the application files require a transportation map, detailing which aspects are existing, renovated or improved, new or planned irrespective of the Games, and new or put in place for the Games (Bovy 2011). In particular, linkages of main Olympic venues to high performance transport, the quantity of transport modes for Games operations, and clustering of venues are highly important. To minimize congestion and travel time for athletes, the IOC sets forth a preferred layout of the competition and non-competition venues: total dispersion or full concentration is to be avoided. Dispersed venues create heavy and expensive logistics between many venues while clustered venues put high pressure on the transportation system, because millions arrive and depart from the same area. From a transportation point of view, the IOC believes that the optimum lies between the two (IOC 2000: 46).

During the applicant city seminar of the 2012 Olympics, Bovy and Protopsaltis, key figures for Games transport operations, laid out the risks and benefits for concentrating or dispersing Olympic venues (Bovy and Protopsaltis 2003: 45). The concentration of Olympic venues downtown or in clusters is a major advantage if transport services offer high frequencies and high capacities. At the same time, the risk of a downtown location and Olympic clusters is transport congestion and more difficult pedestrian crowd management. Comparatively, for outlying, stand-alone venues access design and operations are simpler, but they bear inconveniences for all client groups due to much longer travel times and higher costs of transport services.

Even though both of the Olympic transport experts define risks and opportunities for certain accessibility arrangements, IOC mobility preferences effectively limit the land-use choice of cities. In sum, stated preferences encourage cities to cluster: the Olympic Organizing Committees (OCOG) must design a system to minimize distances and travel times for the Olympic Family, especially athlete travel times shall be less than 45 minutes (Bovy 2004a: 19). The Olympic village and very large 24 hour/day traffic generators like the Main Press Center (MPC) and International Broadcast Center (IBC) should be close to the Games' center of gravity (Bovy 2004a: 26). Implicitly, the rule is: the lesser the travel time for Olympically prioritized client groups, the better for the city's bid.

Olympic transport groups and their mobility requirements

Besides the accessibility requirements for and in between venues, the IOC requires cities to transport their Olympic client groups in a specific manner guaranteeing free flow travel conditions (IOC 1992: 44). Large user categories, such as spectators, must use public transport and more logistically sensitive client groups have to use private transport.

According to the Olympic priorities, the different Olympic transport services are labeled T1 through T5, each associated with different vehicles and service levels.

The following explicit requirements are necessary, because of security, equipment needs and speed of access to the competition venues (Bovy 2004a, 2004b).

VIPs, International Federations, International Olympic Committee, National Olympic Committee, Organizing Committee, athletes and team officials, technicians, media and sponsors (T1 through 4) require private transport. There is no flexibility given to Olympic transport organizers on what type of private transport they choose; it has to be cars and buses. This client group requires on demand 24/7 service with high reliability, punctuality and security. Transport for the first four client groups is free: defined in the Olympic Charter, the OCOG has to cover local transport costs of accredited national Olympic delegations, technical officials, the Olympic Family, and representatives from medical commissions, the Olympic museum, world anti-doping agency and so forth. The T5 transport service group, comprising the staff, volunteers and spectators, has to travel by public transit. For mega events "only very high performance public transport has the necessary capacity to meet the very peaked and high traffic demands" (Bovy 2004b: 20). The transport consultant Professor Bovy, gets even more specific in suggesting that the spectator traffic shall only be supplied by high capacity *rail* transport, because it can deliver traffic demand peaks of 50,000–200,000 persons/hour. Light rail and medium capacity subway help "but are often not sufficient for mega events" (Bovy 2004b: 23). Furthermore, parking close to venues for the T5 client group is prohibited (Bovy 2004b: 22).

In order to coordinate the travel flows, a single entity for transport is needed to oversee all traffic functions. A dedicated transport representative must take responsibility for all internal and external communications via all media outlets (Bovy 2004a: 56). The traffic command, control and communication center is essential for the games-time transport system. It allows efficient and real-time transport and security management.

IOC planning recommendations for sustainable legacies

Having provided the cities with a long list of accessibility and mobility requirements, the IOC aims to support them in preparing for such a massive transport task via further planning recommendations. The first step for host cities is realizing that the IOC requirements cannot be fulfilled by most existing transport infrastructure and regular operations, as "mega event operations often imply major public investments in transport infrastructures and mobility equipments: compressing 25 years of projects in 5 years" (Bovy 2004b: 52–3). Therefore, the IOC places a strong emphasis on "sustainable legacies" aiming to improve the cities' transport system for post-Games travel. Therefore, the Olympic Charter in force since 1 September 2004 has as its fourteenth mission statement to promote "a positive legacy from the Olympic Games to the host cities and host countries."

However, what exactly legacies are, is highly debated among the IOC members and host cities. David Richmond, the head of the Sydney Olympic

Organizing Committee (SOCOG), implied the subjectivity of the matter: "legacy like beauty is in the eye of the beholder" (Richmond 2003: 3). Implicitly, he suggested, sustainable legacies are positive and can include venue infrastructure, expertise and experience, and overall major legacies should not be "add-ons" but integral to the bid. Richmond's (2003: 22) recommendation was to "align and integrate Games planning with 'big picture' city planning and development." The golden question, which remains to be answered, is how?

The IOC and associated city officials provide planning recommendations to help achieve sustainable legacies. Recommendations made by the transport consultant Bovy emphasize long-range planning goals for local communities (Bovy 2004b: 52–3):

- "Mega event infrastructure must be planned and designed in priority to serve long-range community development goals."
- "Long-term components are to be cast 'in concrete' and 'mega event over-lays' built as temporary structures."
- "Mega event transport infrastructure especially heavy rail, motorways, and tunnel projects must be primarily conceived for long-range community benefits and metropolitan sustainable development."

These goals can be achieved by "integrating legacy projects in Olympic plans in parallel with examining each project in terms of sustainable development" (Cartalis 2003: 20–1). Overall these recommendations cover the basics, but are insufficient to actually plan for sustainable legacies.

Olympic legacies: a win–win situation for cities?

The IOC gives the answer to the question "which outcomes should we expect from recommendations?" upfront. By claiming a variety of Olympic benefits, the IOC further entices cities to buy into the Olympic idea. The results for urban development, and in particular for transportation, have remained unquestioned so far. Professor Bovy stated in a 2010 interview: "In 30 years of Olympic experience, I have seen quite a few sports venues turned White Elephants, but I have never seen real Transport White Elephants" (Ostertag 2010: 16–17).

According to the IOC, the Games carry a strong momentum for development and can act as a powerful catalyst for urban projects. Because of this opportunity, most host cities have used the catalytic momentum to substantially upgrade their transport system and built a metropolitan-wide legacy (Bovy and Protopsaltis 2003: 27; Cartalis 2003: 2). For transport, Olympic legacies are generally positive (see also Bovy 2004a: 59). Hosting cities can expect the following specific legacies (Bovy and Protopsaltis 2003):

- integrated public transport system expansion (new or improved transport systems as well as reserved bus lane networks),

- Games' specific transport improvements (public transport and highways) for long-term metropolitan and community needs,
- Games' innovative transport and traffic-management concepts useful for other events and commuters,
- adoption of improved traffic incident or accident management techniques.

Recognizing the need to actually evaluate legacies and Games' outcomes, the IOC launched the Olympic Games Global Impact (OGGI) project in 2001 with the goal to achieve sustainable development economically, environmentally and socially. The program required using 150 indicators during the bid, candidacy and preparation efforts of the city up to two years after the Olympics. Evaluations of these measures comprise the final obligatory deliverable for Olympic host cities. The problem with these measures is three-fold. First, they did not include any qualitative measures, which are hugely important for a holistic assessment of the Games (Leonardsen 2007; Pyo et al. 1988). Second, the evaluation stretches only two years after the Games, while impacts can last for decades or even centuries. Third, no city is accountable for unfulfilled or unsustainable legacy outcomes.

The IOC's power in the planning process

The IOC also has the authority to implement its prescribed transport requirements and shape its Olympic transport vision in host cities. During the bidding and candidacy stage, the IOC apparently holds the most power because of the forthcoming evaluation process. "During the bidding stage, the cities face the most pressure, because they are in it to win" presumes the IOC's operations director in an interview. In the bidding file, the city puts forth its strongest arguments on why it is the best candidate to host the event. This may mean promising infrastructure and operational measures that ensure superior travel conditions for athletes and millions of visitors. These new projects and preparations are crucial in winning the bid: Bovy and Protopsaltis pointed out at the 2012 candidate city seminar that "upgrading and development of new transport infrastructure plays a key role in an Olympic bid and its evaluation" (Bovy and Protopsaltis 2003: 27). Once the city wins the bid, it is responsible for fulfilling the promises made during the application process. In the seven years following a successful bid, the future host city is obligated to build the promised infrastructure and set up the necessary services, as the bid turns into a contract for the winner.

The following sections highlight the IOC tools under the umbrella of the TOK (Transfer of Knowledge) program that can potentially influence a host city's transport and venue location choices during the Olympic preparation process. The TOK program is the primary means of communication between the IOC and cities (Figure 2.1). Through the TOK program, the IOC intends to transfer knowledge and lessons learned about staging the Olympic Games from previous to future hosts (Bovy 2004a). For example, successful planning approaches from the Games organizer (London 2012 monitoring of Beijing 2008) are a strong input into Games bidders' preparations (candidate cities for 2016). Officially, the monitoring program is to diminish potential risks in the conduct of the Games and to search for contingency solutions, should problems arise during the preparation process (Bovy 2011). However, it also helps to ensure cities stay on track with their preparations and keep the promises made in the bid file. While intended to transfer best practices and avoid pitfalls from previous Games, it actively influences future hosts in their decision-making about urban development throughout the bidding, candidacy and preparation stages. While during the bid and candidacy, the cities focus on the IOC requirements and recommended revisions to their bid, during the preparation stage, the IOC still has a final card to play. As a threat – which the IOC has never actually executed – the IOC can take away the Games at any time from the host and give it to the previous Olympic host city in case obligations are not being fulfilled. In Olympic lingo, this is the "red card." Cities receive a "yellow card" if the IOC feels the staging of the Games is in jeopardy.

The TOK program has multiple stages, to which the IOC assigns different priorities (Figure 2.1). The applicant city seminar allows the IOC to inform bidding cities about successful bid factors and to assess the potential of bidding cities to stage successful Olympic Games. To reflect the IOC's priorities, weights are assigned to the different Olympic Games functions in the bidding file: the highest weighting factors are given to general infrastructure (including transport) and accommodation, closely followed by sports venues and the Olympic village, followed by transport operations, safety, security, the general concept and projected legacy (Bovy 2005). Bidding cities are trying to at least meet if not exceed the set IOC benchmark; for each Olympic function they have to score at least 6 out of 10 points (Bovy 2011). The second bidding stage – the candidacy – allows refinement of the bidding file and implementing changes the IOC had suggested (Bovy 2011). Once the Games are awarded, several further tools can influence the host cities development process. For example, Olympic Games Knowledge Management (OGKM) seminars and more than 40 Olympic Function Technical Manuals help cities prepare for the Games, while they can always seek IOC advice on handling the Olympic challenge. With a clear timeline in mind, seminars and personal advice to candidate cities paralleled by frequent monitoring ensure the cities meet their promised targets. In detail, these include:

- The Evaluation of Mandatory Questionnaires: As part of the bidding process, questionnaires are distributed by the IOC (IOC 2004). The initial questionnaire has a specific transport section, covered in Theme 14, that requests information about existing, planned and additional transport infrastructure, the main parking areas, fleets for different travel modes, distance and journey times (between the competition venues as well as the training venues) for the application year and the year the city is hosting the Games. The goal of the

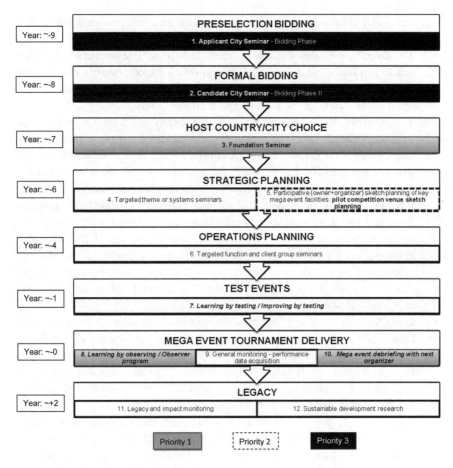

FIGURE 2.1 Transfer of Knowledge Program

Source: Created by author. Adapted from P. Bovy, IOC Transport Consultant, personal communication via email 30 July 2008.

questions is to provide the IOC with an overview of the candidate city's transport network and operational plans for the Olympic Games (Bovy 2005). Selected candidate cities can use the IOC "acceptance evaluation to orient, adjust, and improve their bid quality and coherence" (Bovy 2005).

• The Pre-Selection Visit: Before selecting a city as a host, the official evaluation commission of the IOC arranges a full pre-visit briefing and analysis of the whole candidacy dossier. This visit includes two days of "in-house" comprehensive presentations of all bid themes, several days of visits of all proposed competition and non-competition venues, and a full wrap-up session and evaluation commission press conference. This pre-selection visit concludes with a qualitative risk assessment of all bid themes based on the initial bid report and complementary information provided during this official IOC visit.

- The In-House Evaluation and Informal Inquiries: The option for in-house analysis of technical themes by IOC experts (security, transport, economic development, etc.) always exists. These experts are available for confidential inquiries throughout the bidding phase (Bovy 2005).

- The Seminars and Supporting Documentation on how to prepare the bid: Since 2002, the IOC organizes a four-day acceptance of candidate cities seminar on Games structure, content, main organizational functions, past most successful experiences and key transfer of knowledge lessons. Furthermore, candidate cities receive large amounts of material from the IOC on how to bid. These include the Transfer of Knowledge material from previous Games as well as mandatory preparation details. The documents can amount to six 650-page dossiers including maps, CDs and evaluation reports (Bovy 2005). In the past, cities have readily accepted these recommendations. Because the two-step bidding process and substantial Transfer of Knowledge Program, the IOC has received high quality "competitive" bids (Bovy 2005).

- Live Observer Program: City planners for future Olympics participate in the Live Observer Programs of an Olympic host city, because the IOC believes that training the core staff through participation in preceding Games is key to the success of future Games (Bovy 2004a: 27). Critical monitoring functions include sports management, transport, security, accommodation, technology, media, arrivals, departures and so on.

- Approval of Transportation Plans: For transport, the most central concern is the establishment of the Olympic Transport Strategic and Business Plan. Cities have to submit this plan for approval to the IOC (44 months prior to the Games). Final transport plan approval occurs 12 months prior to the Games. All changes to the original bid or these plans have to be approved in writing by the IOC.

- Monitoring of the Transport Implementation Process: Two IOC monitoring streams for future host cities exist: M1 and M2. M1 monitors the promised infrastructure and its implementation, which receives its final update four and a half years prior to the Games. M2 monitors the operational systems, including testing events at least one year prior to the Games.

Legacies, in this TOK's hierarchy, are of minor priority (Figure 2.1). By 2011, the TOK priority assignment had remained unchanged (Bovy 2011). A potential pitfall of the knowledge transfer is that a critical review of existing measures is neglected. Furthermore, tactics applied in one city might not be the best choice for other cities given the socio-economic or environmental circumstances. Finally, unsustainable practices can be passed from one city to another, especially if the evaluation period of the Olympic Games Global Impact (OGGI) program extends only two years after the Olympic Games.

Conclusions

To ensure that the city is prepared for the Olympic transport challenge, the IOC exercises a strong influence over hosting cities during the planning process. Cities frequently comply and fine-tune their preparations motivated to stage the perfect transport Olympics. Through the setting of objectives, demanding of priorities and putting a strong weight on transport during the evaluation process, the IOC has a strong influence on Olympic transport, and hence its legacies. This influence lasts not only during the two years leading up to the election of the host, but also during the implementation stages. Therefore, the challenge for aspiring Olympic host cities lies in aligning the short-term needs for the event and the long-term strategy for city development.

If host cities fail to align the potentially competing interests, legacies can be detrimental to the host. Even though the host city and the IOC largely share the goal of making the visitor experience an excellent one, the city has broader goals. They not only include transport services and infrastructure to meet event needs, but also long-term development objectives. Getz (1999: 24) hypothesized that tradeoff decisions generally tip in favor of the Olympic short-term needs: "the event organization will likely be dominated by people whose mission is to produce the event on time and within budget so the task of planning and monitoring related effects is left to the government or 'watchdog' groups." For the city there is a clear tradeoff between short-term service delivery and cost. The IOC does not feel this tradeoff, because it only pays a comparatively small amount; thus, it is likely to demand more and better services than the city is likely to want to provide and pay for. The city also may or may not find tradeoffs between short-term service delivery for the event and its long-term development strategy. It is detrimental when the desire to win the bid leads cities to promise exceptional alterations to urban areas that are not in their long-term interests. Because once the bid is won, host cities are locked into their promises through the host city contract.

Olympic legacy doubts

Planning recommendations and legacy promises made by the IOC have stimulated expectations about what the Olympics can bring to a city. As early as 1988, Pyo concluded that "above all, the Olympic Games should be recognized as an investment for the future and an image building event rather than a profit generating opportunity" (Pyo et al. 1988: 144). Legacy expectations have gained a strong momentum among bidding cities and are advertised positively prior to a bid. Common newspaper headlines in the run-up to the Olympics, like "mass transit crisis could complicate city's Olympic prospects" (Hersh 29 May 2008), promise massive improvement in the public transport systems. Are the promises made and possibilities for sustainable legacies offered by the Olympics to hosting cities realistic?

There are some indications that they are, and some indications that they are not. For example, Bovy concludes that "mega events are a catalyst for main metropolitan transport infrastructure developments, which are needed but would otherwise have been constructed in a much longer time frame" (Bovy 2004b: 45). Those academics, who have challenged general promises on urban development, express doubts about the purely positive nature of legacies. "In their enthusiasm to be chosen as host city in competition with other bid cities, many cities succumb to starting huge projects" (Preuss 2004: 79). As a consequence, these projects can be counterproductive to the city's long-term development plans.

The IOC paints a positive picture for potential bidders, leaving potential risks or failed Olympic legacies often unmentioned. For example, Cartalis (2003) pointed out that the Schinias rowing center, built during the Athens Olympics, was a good example of sustainable planning. Controversially, local media reports indicated the opposite. They reported that the Schinias rowing center destroyed the natural habitat for many species and was a "disaster" for the environment and historical site (Pappas 2003).

Whether these claims by academics, the media and the IOC are actually substantiated or not, has yet to be proven. Long-term evaluations of impacts of mega events barely exist, and, if they do, they are primarily descriptive. The key question is whether, given Olympic priorities and requirements, it is reasonable to assume or even possible to accomplish true sustainable legacies in cities.

3

PLANNING FOR THE 1992 OLYMPICS AND BARCELONA'S URBAN LEGACY

The summer of 1992 marked the beginning of a magnificent rise to the world stage for Barcelona, when the Olympic Summer Games transformed the city's urban structure, its neighborhoods and its transport system. Since then, the city's urban transformation has become an inspirational model for other Olympic hosts, setting the highest standards on how to use the Olympic Games as a stimulator for urban development and competitiveness in the global realm. This transformative scheme became famously known as the "Barcelona model" (Capel 2007), in which the Olympic catalyst expedited Barcelona's development by completing key projects within five years that would have taken 30 years without the games (Mosby 1992). The story of Barcelona's success, however, is deeply rooted in a careful planning process for Olympic legacies. The following tale sheds light on a multitude of planning facets stretching from political engagement, personal relationships and strong individuals, all of which brought the politically neglected yet historically rich city back to its path on capturing the imagination of architects and allowing the city to rightfully claim the world stage.

Barcelona's urban and transport history

Barcelona, the capital of Cataluña, is located in the southeast of Spain on the Mediterranean Sea. Under Spain's dictator Francisco Franco, Barcelona and Cataluña experienced 40 years of neglect with little investment in public amenities. During this time the city grew mostly uncontrolled, only constrained by natural boundaries: the sea, the rivers Llobregat and Besòs and the Serra de Collserola ridge (Sànchez et al. 2007). During Barcelona's development years (between 1960 and 1970) the city experienced rapid industrial and economic growth, accompanied by urban chaos and a lack of planning. After Franco's death in 1975, Barcelona was left with both a legacy of economic and social isolation, and the problems of its recent rapid urban growth. To make things worse, Spain was stricken by a severe social and economic crisis marked by inflation, a declining population and high unemployment.

The neglect of investment in public infrastructure was also apparent in Barcelona's transportation system (CMB 1986). While the public transport usage had been in steep decline, motorists pushed onto the city's already clogged road network and significantly increased air pollution in Barcelona's center. After a democratic city government was elected in 1979, two entities for transport were established by the Catalonian parliament to remedy the transport problem: in 1986, the Center Metropolitan d'Informacio i Promocio del Transport S.A. (CETRAMSA), the metropolitan transport company; and in 1987, the Entitat Metropolitana del Transport (EMT). EMT was a coordinating committee formed by 18 municipalities in the Barcelona metropolitan area with the task to provide joint public passenger transport services. The rail transport system was operated by three rail carriers: renfe, the Spanish railway company, the metro and the regional railway system managed by the Ferrocarrils de la Generalitat de Catalunya (FGC). Bus systems were operated under the umbrella of the Transports Metropolitans de Barcelona (TMB) (TMB 2008: 4).

After years of neglect, Barcelona sought a holistic metropolitan strategy to escape its urban crises. The goal was to emerge as a global city with a revitalized Catalan identity and strong political leadership (Borja 2004). For urban regeneration, the Ajuntament de Barcelona (Barcelona's city government) sought to solve the city's three main problems: the degradation of the city center, the lack of urban consolidation of Barcelona's outskirts, and the need for rejuvenation of the sea front. Therefore, the city government established aspirational goals that were to guide the future development of the city: foster regeneration of Barcelona and its individual city quarters, open a discussion of "grand schemes of change" on the metropolitan scale, and develop metropolitan services such as transport, public spaces and so on, which would satisfy the needs of the citizens (Argilaga and Benito 1997). To realize this scheme of urban revival, there were three distinct plans (the latter two had to be modified due to Barcelona winning the Olympic bid):

1. The special plans for Interior Reforms (PERI = Plan Especial de Reforma Interior), focusing on the uniqueness of different neighborhoods, was a means of channeling citizens' demands for public spaces and facilities and the conservation of historical buildings that were not covered by the following two plans.
2. Plan for the New Center Areas (ANC = Arees de Nova Centralitat),which selected 10 undeveloped areas that were to receive a concentration of services and facilities making them new focal points in the city. The locations of these 10 areas are marked in Map 3.1.
3. Plan General de Metropolitano (PGM) of 1976, which included the development of Barcelona and its 26 surrounding municipal areas laying out "la reconstrucción de la ciudad" (the reconstruction of the city). The

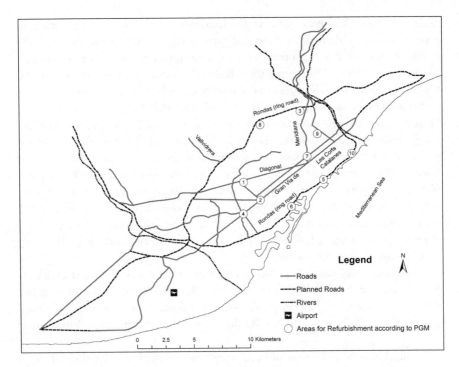

MAP 3.1   Plan General Metropolitano (PGM) 1976

Created by author with GIS (country boundary layer) and georeferencing points according to multiple Google maps. Planned roads and existing roads adapted from Guàrdia i Bassols et al. 1994: 89 (further modified by author in black and white). Areas for refurbishment adapted from Sànchez et al. 2007: 15 (this author reproduced from Esteban 1999)

PGM sought to establish a new urban order through a clear layout of public spaces and a road system that was to prevent further congestion in the urban network (Map 3.1).

Barcelona's candidature and Olympic preparations

In the socio-economic decline of Barcelona, the Olympic candidature became a collective dream guided by the city's metropolitan vision. Nationwide, Barcelona experienced political stability and continuity in the run-up to the Olympics. Among all political parties, there existed a strong sense of agreement regarding the Olympic dream of revitalizing the city (Interview Sala-Schnorkowski 2007). Interestingly, after the Games, a lack of political stability has stalled further rapid transport and urban investments in Barcelona (Interview Pàmies Jaume 2007).

Barcelona's Olympic aspirations first surfaced in 1980, when the mayor, Narcís Serra, discussed the idea of Barcelona hosting the 1992 Olympics with Juan Antonio Samaranch, the IOC president at the time. In 1981, five years before the official IOC application was due (1986), Serra officially announced Barcelona's desire for the Olympics, which the Barcelona City Council unanimously supported shortly thereafter. Plans and early investigations of the bid (e.g., the Cuyas Report of 1982) contributed to a preliminary report about a potential candidature (1983). Following the planners' idea to take advantage of the Olympic opportunity by regenerating Barcelona's urban areas, the famous architect and Olympic lead planner for Barcelona, Lluis Millet (Interview 2007), said: "It is not good that the politicians are involved from the beginning [in the planning]. First, it had to be the planners, the masterminds. It is not a political problem. It is a technical one." His statement suggests two important planning features that contributed to Barcelona's successful redevelopment. First, good planning not political motives should drive the alignment between the metropolitan vision and the Olympic requirements. Second, planners have to guard their plans against political influences. In these critical planning stages, Millet (Interview 2007) and a few other architects prepared plans, supported by "very strong relations with the mayor."

In the following years, Barcelona secured support from Madrid and started its preparations for the Games by giving presentations to the IOC and forming the Barcelona Candidature Governing Council. Barcelona's city council quickly realized the strong influence of the IOC in the wake of the Olympics on urban politics.

> In this sense, the institutional consensus created during the Olympic Games broke with a negative inertia in the way of doing things in Barcelona, characterized basically by the fact that traditionally, it was only the municipality which confronted itself with urbanistic and sectoral policies, which, because of their nature had to be the object of shared attention with the central and autonomous [Catalonian] governments.
>
> *(Acebillo Marin 1992: 25–6)*

The election of Barcelona as the host of the 1992 Olympics took place in Lausanne on 16 October 1986. On 12 March 1987, COOB'92 (Olympic Organizing Committee for Barcelona '92) was officially constituted as the supreme governing body of the Olympic Games. Even before the IOC officially elected Barcelona as the host of the '92 Olympics, the city had begun to enlist volunteers and had started construction (NBCSports 1991).

Throughout Barcelona's candidacy and preparations, Olympic planners faced extra scrutiny. Samaranch, born and raised in Barcelona, was the IOC's president at the time of Barcelona's bid, candidature and victory. He communicated intensively with the planners and influenced Barcelona's planning process.

> Samaranch was in our favor. He was our worst enemy. And we had to do it much, much, much better, because he was our worst enemy. He was our worst critic: "this one no . . . this has to be that . . . this" [quoting Samaranch]. We had to do it much better. We knew more than them (the other candidate cities). We were talking with everybody and we

were always better than what other cities were presenting. It was the only way to win. Today, Barcelona would not win, impossible. I am very clear about this.

(Interview Millet 2007)

How much Samaranch's personal interest and existing relationships influenced Barcelona's election as the 1992 hosting city is unknown. It was important, however, that Samaranch took a personal interest in the host city. thereby guiding Barcelona's comprehensive future development. Overall Millet (Interview 2007) says the IOC itself – not their requirements – barely influenced the planning work of the Barcelona Organizing Committee, because at the time Barcelona was hosting the Olympics, the IOC did not have the necessary expertise. In fact, according to him, Barcelona was the foundation stone for future expertise used.

Upon the award to stage the Games in 1992, the Olympic preparation needs led to an immediate change in priorities: "The right to organize the 1992 Olympics . . . evoked at first an enormous wave of enthusiasm and satisfaction, and immediately a universal reorganization of priorities of all kind" (Barreiro et al. 1993: 20). These new Olympic priorities tremendously affected Barcelona's urban projects and ultimately resulted in a modified master plan and new urban structure: four areas within Barcelona would serve as centers of Olympic activity, which – combined – fulfilled what was necessary to stage the Games. These areas were chosen due to existing and planned sporting facilities: 37 were required, 27 were already built, five were under construction (1985–6), and the other five were in the planning stages (Table 3.1).

Note the change in priorities and directions of urban planning that is visible when the Olympic layout (Map 3.2) is compared with the areas for refurbishment according to the PGM (Map 3.1): the Olympic areas of Vall d'Hebron (corresponding to 8 in the PGM) and Parc de Mar (corresponding to 5 in the PGM) were included in the initial ANC plan, while the areas of Montjuïc and Diagonal were added by way of the Olympics as areas of refurbishment (CMB 1976). This set-up clustered sports venues and contained most of the athlete's and IOC's accommodations within a circle of 5km, in which 85 percent of the sporting activities would take place, thereby minimizing travel for athletes and visitors alike. This change in priorities due to the Olympics – per Millet (Interview 2007) – was also beneficial for Barcelona: "the city needed a structure

TABLE 3.1 Barcelona's four Olympic areas

	(1) Montjuïc	*(2) Diagonal*	*(3) Vall d'Hebron*	*(4) Parc de Mar*
Area in Map 3.2	southwest	northwest	northeast	southeast
# of venues	5	6	5	2
# of sports	13	7	5	4

Sources: Compiled by author. Adapted from NBCSports 1991; BOCOG 1993.

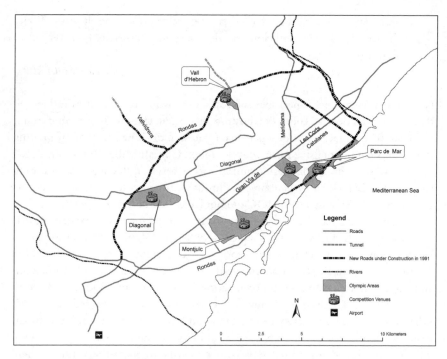

MAP 3.2 Barcelona's road construction projects (1991–2)

Created by author with GIS (country boundary layer) and georeferencing points according to multiple Google maps. Road construction and Olympic areas adapted from Holsa 1992b

to accommodate its future growth. We put an Olympic area at each entrance gate of the city. It was easy, a clear vision."

The four Olympic centers of activity

The mountain of Montjuïc was the heart of the Barcelona Games. Located 70m above the city with the Olympic Park and 13 allocated Olympic sports, it contained more venues than the other Olympic areas. The area around Montjuïc required significant infrastructure investments to integrate the beautiful land-scape back into the city. The Montjuïc area has always had serious accessibility problems from the city (Holsa 1992a). Before the Games, the primary means of access to Montjuïc was by foot. Given that Montjuïc was expected to welcome close to 2 million spectators and thousands of athletes for the Olympics, several access alternatives were considered. For the Barcelona Organizing Committee for the Olympic Games (BOCOG), it was clear that private cars had to be constrained, but the capacity that could be provided with taxis, shuttles and buses alone was insufficient. After long political battles and conflicts within govern-mental organizations, visitors to the Olympics were provided with three alterna-tive modes of transport to Montjuïc:

1. *a bus shuttle service* from the Plaza España to the Olympic area,
2. *mechanical escalators* embedded in the slopes of the mountain. Specifically for the Games, three new escalators from the Plaza de España to the Olympic Ring were installed, permitting access to the sport venues within 15 minutes (Holsa 1992a). This system was considered to function well during the Olympics because of its high flexibility and reliability.
3. *the funiculars*, a mountain train from the Parallel metro station. This option was expected to attract the most passengers. Due to the International exposition in 1929, the funiculars already existed, but the Olympic Games required the complete refurbishment of the mountain train (Sala-Schnorkowski 2007). After refurbishing the old system during 1991, the funiculars carried about twice as many passengers yearly, most of them tourists (Ajuntament de Barcelona 1993, 1995, 1999).

The fourth access alternative to Montjuïc discussed among the decision makers was the refurbishment of an old, since the 1980s abandoned metro line (CEAM 1987: 88). It was to be extended from St. Antoni underneath the Montjuïc Mountain to Barcelona's airport El Prat including several stations.

In the end, the political focus on the sufficiency of existing metro lines and the cheap alternative of the escalators available overrode any discussion about revitalizing the abandoned metro line prior to the Olympics. The extension was not implemented, even though both the mayor and the public transport committee supported the metro's refurbishment. Their arguments were powerful: the main section of the line already existed and forecasts predicted 45.8 million passengers yearly post Olympics (CEAM 1987; TMB 1988). The TMB president at the time, Merce Sala-Schorkowski, asked the IOC to support the metro's extension, but was told by IOC's president Samaranch (1989) that the time frame for implementing the project was simply too short. So went the official explanation. Unofficially, the metro expansion funds had been tied up through "more important" Olympic projects (Interview Millet 2007). Because the Olympic Montjuïc station would not generate enough Olympic demand (TMB 1990) and other – cheaper – alternatives were available, the metro line received lesser priority among the possible Olympic projects. Revitalization of the metro line started eventually in 1995 and was completed within two years (Mateu Cromo 1996; Schwandl 2004).

The second Olympic area, the Diagonal, was one of Barcelona's upscale areas with an established university, high-class hotels (lodgings of the IOC members) and expensive shops (NBCSports 1991; Sànchez et al. 2007). The third Olympic area, Vall d'Hebron, used to be a remote neighborhood with a middle-class residential area (NBCSports 1991). Residences for the referees, judges and media were added by way of the Olympics (Barreiro et al. 1993). Transport connections to both areas were excellent, thanks to all existing metro lines and the newly built ring roads.

Placing the Olympic Village in the run-down area of Poblenou, to be called the Parc de Mar Olympic area, was the key decision to inaugurate the recovery of the sea front and make Barcelona's metropolitan vision become reality. Parc de Mar, one of the most traditional districts of Barcelona, was the location of old factories, workshops and many abandoned buildings. In designating this area as the home to the majority of the 15,000 athletes, it had to have a pleasant atmosphere, and guarantee optimum security (Martorell et al. 1991). Therefore, the Ajuntament de Barcelona undertook vast improvements to the area including a metro station, a Sport port, beaches and new sewage systems. Furthermore, 2,500 new stores, new residential complexes, hotels and two high-rises for new offices were built in the run-up to the Games (Barreiro et al. 1993). All these leisure facilities combined, turned the former industrial area into a center of urban life, which remained a lasting urban legacy. Eventually, it became evident that the Olympic Village was no place for families. Because schools and kindergartens had not been provided for, the area did not attract any families (Interview Brunet 2007).

The cooperation among public entities to transform the sea front, so Sala-Schnorkowski (Interview 2007) believes, was only possible because of the Games and very influential people involved in the process. For example, the two renfe train tracks running along Barcelona's coast had to be removed. The problem of the train tracks blocking access to the sea and thus hindering the reurbanization of Barcelona was already evident in 1981. But not until February 1986 – 18 months before Barcelona was elected as the 1992 host city – an agreement between the president of renfe, the Ajuntament de Barcelona, the Ministry of Transport, Tourism and Communication, and the Metropolitan Corporation could be reached, announcing the plan to remove the coastal tracks and transform the Franca train station (Delegacion de la 5a zona 1989). The Franca train station was later abandoned, remaining only as a historic building. To compensate for the removal of the existing train tracks, a formerly freight-only line was upgraded to allow suburban trains access to the city center (CEAM 1987). After the train tracks were removed, the Ajuntament de Barcelona was able to create six artificial beaches (San Sebastia, Barceloneta, Nova Icario, Bogatell, Mar Bella and Nova Mar Bella) and a new port, where Olympic regattas could be held (Ajuntament de Barcelona 1993; Holsa 1992b). After the Olympics ended, the unique cooperation that had been in existence for the Olympics dissolved (Interview Sala-Schnorkowski 2007). Hence, the Olympic Games provided the catalytic momentum in discursive and spatial terms to create Barcelona's revolutionary development plans.

Transport changes due to four Olympic centers of activity

With the change in priorities for areas of urban refurbishment and the selection of four Olympic centers, regional transport investments also received different priorities in the run-up to the Olympic Games. With the dawn of the Olympics, Mayor Serra announced an additional goal specifically for transport: to connect Barcelona to other cities in Cataluña through regional transportation systems. In reality, these were primarily road projects because of the Olympic influence. Three aspects support this observation. First, the ring road circling Barcelona's center provided frequent and unhindered travel in between the four Olympic areas (see Map 3.2). Second, the Vallvidrera tunnel provided access to northern venues (shown to the north of Barcelona's center, Map 3.1). Third, the fact that no major public transit investments were made prior to the Games, because Olympic spectator movements could be accommodated by the existing metro lines that efficiently linked all Olympic areas.

For Barcelona, the ring road "was an absolute necessity" said Oriol Pàmies Jaume (Interview 2007), the press contact of TMB. Each Olympic area was adjacent to these newly constructed roads (Map 3.2), which provided the essential connection between the four Olympic areas: Montjuïc, Parc de Mar, Vall d'Hebron and Diagonal (Holsa 1992b). The Games catalyzed the ring, guaranteeing the completion of the circuit before 1992 (Millet 1995). Seemingly, giving priority to the road infrastructure was in the interest of the IOC. As Merce Sala-Schnorkowski (Interview 2007), the CEO of TMB during the Olympics, remembers: "the IOC's priority was to easily transport the Olympic Family members and hence improve the transport networks of the roads and streets of the inner city." Given that all the VIPs and athletes were being transported by car or bus, building the ring road was an important consideration in the planning stages. A long-standing reporter for Barcelona concurs that the Olympics changed transportation plans in Barcelona. "A drastic rethink of the planned road network led to the adoption of a much more ambitious scheme, involving the construction of a 21 km outer ring system" (Cross 1992: 12).

The reasons were many to go ahead with the road project at the time of the Olympics. Since 1963, advocates for this ring road had stressed the city's inadequate capacity to satisfy current and future travel demand between the region and the metropolis (Barreiro et al. 1993: 9–11). The transport and traffic area coordinator, Alfred Morales Gonzalez, estimates that it would require increasing the capacity of the metropolitan region from the current 588,000 to 900,000 vehicles per day to satisfy predicted travel demand. Otherwise, the road system would reach saturation in 1995 (Servicios Tecnicos de la CMB 1993).

In the city's center three main access roads, the Diagonal, Meridiana (north–south) and the Gran Via (east–west) intersected, causing heavy congestion (Map 3.2). Mayor Serra hoped to relieve congestion for the future growth of the city via these new ring roads. Again technical studies supported this road scheme. Building the ring would result in a significant drop in inner-city traffic, causing a 15 percent reduction of traffic density and reducing travel time more than 90 percent (Servicios Tecnicos de la CMB 1993). Furthermore, the ring road would reduce noise and pollution levels and enhance access to the airport (Preuss 2004). However, these planners also acknowledged that the new road would relieve congestion only for a few years (Riera 1993).

A further change to Barcelona's regional road plans was the expansion of the Vallvidrera tunnel (Map 3.2). While it was needed for the Olympics, the tunnel became the basic infrastructure connecting Barcelona to the city's northern suburbs, Terrassa and Collserola (Holsa 1992a). At the same time, the completion of the tunnel was "of crucial importance for the celebration of the Olympic Games" (CEAM 1987: 82). To make the Vallvidrera tunnel a viable travel route, five smaller tunnels and further roads had to be built.

The powerful Olympic stimulant erased many and varied problems that had hindered building the roads earlier on (Servicios Tecnicos de la CMB 1993). Since the 1970s the ring road had encountered fierce resistance from the residents of the neighborhood adjacent to the ring. The residents opposed the idea of having a large highway cut through their city and demanded instead that additional metro lines should be installed (Sànchez et al. 2007). While Barcelona's regional road network expansion seemed in retrospect perfectly aligned with Barcelona's urban vision, only the ring road surrounding the central city had been in the 1976 PGM (Map 3.1). The expansion of the Vallvidrera tunnel was never part of Barcelona's expansion plans and was added by way of Olympic needs. However, plans for the Vallvidrera tunnel did appear in a previous master plan of the city, the Plan Comarcal of 1953 (Guàrdia i Bassols et al. 1994).

Besides these road improvements and the access options to individual Olympic areas, Barcelona invested in two further transportation projects prior to the Games: upgrades to Barcelona's airport El Prat and the refurbishment of the Estació del Nord (Northern Station), a former train station. In the wake of the Olympics, the airport received a new international passenger terminal, new aircraft parking platforms and new car parks (Bofill 1993; CIO 1985). Before the Games, the airport was exclusively served by renfe and private motorway access. For the Games, a new airport bus service was introduced in December 1991, which was expected to attract close to a million passengers yearly after the Games (Cross 1992). The Estació del Nord was renovated for the 1992 Olympics. Used as a depot as well as a major hub for Olympic bus services during the Games, it served as the terminus for long-distance and regional bus services after the Games.

Transport performance during the 1992 Olympics

Barcelona hosted the Summer Olympic Games from 25 July to 9 August 1992. Because infrastructural improvements alone were not enough to satisfy the requirements the Olympics placed on Barcelona (ASOIF 1990), the city undertook a wide variety of temporary measures to accommodate the traffic and transit flows. Barcelona's secret to success was "planning for a lot more people than were really coming" as Manuel Villalante (Interview 2007), the former head of the transport operations of the Barcelona Games, judged in retrospect.

For the Olympic Family and IOC members, Barcelona implemented special bus lanes, which ran on exclusive sections of inner-city roads and the new ring road. Cars and buses allowed to travel on these special bus lanes had priority over other public transit services and car traffic (COOB'92 1990). Only during the first day of the Olympic celebration, so Villalante (Inverview 2007) remembers, did "the buses for the Olympic Family have problems with reaching their destinations, because the bus drivers were from all over Spain."

The general public and spectators were encouraged to take public transport, which was free of charge for Olympic ticket holders (TMB 1989). Several public transport routes provided access to the venues, whereas two metro lines carried the largest passenger loads (COOB'92 1990: 36). Besides increased capacity on the existing rail and bus networks, Barcelona's city council provided additional public bus routes to and from the Olympic venues and to central locations in the city (TMB 1992). Overall, Barcelona's bus network was expanded by 20 percent and given the right-of-way through a special traffic light system, experimentally introduced before the Games (COOB'92 1990; Cross 1992). Despite these efforts, the bus systems received harsh critique. On 10 August 1992, Klein (1992) an experienced journalist for the Games, wrote that it was not easy to report on the Games due to the unreliable bus service: "Some scheduled buses did not run on time and some did not exist." Park-and-ride facilities along the rail systems allowed residents and tourists access to the inner city (COOB'92 1990). The base load on Barcelona's road network was reduced through the implementation of traffic demand measures, such as flexible working days and early summer holidays for residents. During the Games, COOB established traffic-control zones around the old city with parking and access restrictions (Cross 1992; Culat 1992). In a tourism survey measuring the Olympic experience of visitors, the access to Olympic areas, the park and rides, as well as the parking restrictions scored below average. In comparison, the Olympic ticket for public transport received very high marks (Brunet 1993: 110). Residents only criticized the noise during the Olympics: "The cars of the Olympic Games jammed the Passeig de Gracia area, their roar nearly drowning out the birds in the trees" (Mosby 1992).

Barcelona's public transit agencies used the opportunity the Olympics presented to modernize their fleet. Buses were fully air-conditioned, the metro received new cars, and renfe bought new suburban electric trains guided by a new electronic passenger information system (Cross 1992). Most of these investments benefited metro Line 3, which carried most of the Olympic load. Also, in the run-up to the Olympics, Intelligent Transportation Systems (ITS) for access control and speed regulation were installed on Barcelona's road system. The ITS were capable of issuing immediate maintenance warnings and were likely to reduce response time in emergencies. The system could also be used to give priority to public transport (Ajuntament de Barcelona 1988, 1989a, 1989b). To coordinate these temporary transport measures, Barcelona built a new technology operations information center. In the long term, Barcelona's city council hoped that these high tech systems would maximize social benefit through time savings and enhance environmental protection through emission reduction.

Transport outcomes and Barcelona's legacy

"The planning focus was on after the Olympic Games" emphasizes Sala-Schnorkowski (Interview 2007). Because of this mindset, a clear vision of

Barcelona's future and strong leaders to guard this vision, Barcelona looked fundamentally different after the Games. In sum, the Olympics provided the impetus for Barcelona's radical transformation into a beautiful Mediterranean city (Botella 1995).

Investments in transport infrastructure during the seven years leading up to the Games, however, were split very unevenly; about 95 percent went into new roads (calculated from Brunet 1993). Officially, these investments were justified by Barcelona's long-standing problem of continuous traffic congestion in the inner city and the lack of capacity to meet traffic demand from the outskirts. Unofficially, and as argued in this Barcelona narrative, the roads were an absolute necessity for the IOC and Olympic athletes, whereas public transit to serve the peak passenger demands of visitors was already sufficient. Sala-Schnorkowski (Interview 2007), the CEO of TMB during the Olympics, emphasizes that the Games brought benefits primarily to road users. Even though the plans to upgrade Barcelona's roads had been around for a long time, the Olympics provided "the necessary drive to implement them" (Interview Sala-Schnorkowski 2007). Building the ring road changed the traffic-flow patterns immediately from centralized axes to a circling ring and as predicted, relieved congestion in the inner city (Ajuntament de Barcelona 1990: 387; 1993: 427). Five years after the Games the Fundació Barcelona Olímpica conducted a survey measuring citizen's satisfaction with the infrastructure built for the Olympics. The Rondas were selected as the most important infrastructural change, ranked even ahead of the Olympic Port, the new sea front, the Olympic Village and the sport installations around Barcelona (Anguera Argilaga and Gomez Benito 1997). All Barcelonan interviewees agreed with the Director General for Land Transport on the fact that "the Rondas were an absolute necessity for the city, from which the residents still profit today" (Interview Villalante 2007). Arguably, because of the Rondas the city's metro ridership declined after the inauguration of the roads between 1991 to 1994 (Ajuntament de Barcelona 1993, 1995, 1999). However, almost two decades later the city continues to face the same challenge of traffic congestion in the inner city and some citizens called into question the decision not to implement a pricing structure for road usage at Games time (Interview Pàmies Jaume, 2007).

In contrast, "public transport did not experience major changes, because it played "a minor role during the Olympics, primarily solving access to Montjuïc" (Interview Sala-Schnorkowski 2007). Some urban projects, as Brunet (1993: 82) summarizes, were not implemented in the run-up to the Olympics. With particular reference to transport, he mentions the metropolitan transportation system and the construction of the high-speed rail line linking Barcelona with Montpellier and Madrid. According to Pàmies Jaume (2007), two beneficial transport initiatives were inaugurated during the Olympic Games, but not continued afterwards. First, an integrated fare system that allowed passengers to switch freely between transport modes allowed efficient travel across the host city. It took Barcelona another 15 years post Games to implement integrated ticketing. Second, accessibility for people with special needs was started with thousands of low-floor buses and the metro was adapted for this population group. By 2007, the process was finally completed (Interview Pàmies Jaume 2007). Furthermore, Pàmies Jaume (Interview 2007) acknowledged that a few parking restrictions and methods (such as the "green zone" for residential parking) have been kept after the Olympics had left.

Conclusions

In Barcelona, the Olympics had a decisive influence on the city's urban and transport-planning process. The decision to add two Olympic areas to Barcelona's original urban plan made further changes to the transport system compulsory to accommodate the movements of athletes and spectators between all Olympic areas. These included adding subway stops, refurbishing the most traveled metro line, overhauling the traffic management system, building the Vallvidrera tunnel, providing escalators and the funiculars – all projects that were not slated to be undertaken prior to winning the Olympic bid.

At the same time, Barcelona also undertook projects that were planned prior to the Olympics and catalyzed them, such as a ring road circling the inner city. Most of the investments in the run-up to the Olympics benefited private rather than public transport. These road investments for the city's future, for example the Rondas, and the IOC's preference for road transport aligned well strategically. Especially the ring road fulfilled Barcelona's transport goal to relieve inner-city congestion and connect Barcelona's suburbs to its center. Conversely, other urban projects – especially in public transport – were delayed because new transport priorities imposed by Olympic requirements (e.g., revitalizing an old metro line). The main investment in public transport – the funiculars – primarily benefits tourists. Removing the train tracks along the coast improved public access to the sea front, but by no means improved accessibility within the city of Barcelona.

Even though the Olympics imposed many requirements on Barcelona, in 1988 – the year Barcelona won the bid – Olympic requirements were comparatively lax. Responsibility for success lay with the city, including learning from previous experiences. As cautionary examples, Barcelona had its two immediate Olympic predecessors – Montreal and Los Angeles. The former ran up a tremendous public debt; the latter staged privately financed Games without public investment. Barcelona's goal was to use the Olympic Games to rejuvenate a city that had been neglected for decades. Strategically and with a very long planning horizon, Barcelona aligned the Olympics in exemplary fashion with its urban vision and transport goals.

4

PLANNING FOR THE 1996 OLYMPICS AND ATLANTA'S URBAN LEGACY

Atlanta missed the Olympic opportunity to bring about revolutionary change to its transport system, even though the city's goal was to evolve out of the Olympic preparations as a "modern multi-modal transport city" (CODA 1993: 3). Instead, the Centennial Games catalyzed primarily improvements for road users. Massive road expansions across Atlanta's counties with high-tech traffic management equipment, high occupancy vehicle (HOV) lanes and a new traffic management center were integrated into a broad Intelligent Transportation Systems (ITS) scheme, which remained in Atlanta as a strong legacy and ultimately driver for its future development. Public transport improvements were relatively minor, including an extension of one rail line and some biking and walking paths. Atlanta's urban legacy included refurbished facilities in the city center, a new inner-city park and new housing developments adjacent to Olympic venues. Because Atlanta's Games were driven by the private sector, whose interest lay in revitalizing downtown for real estate purposes, the bidding committee sought to spend the bare minimum on public infrastructure.

Atlanta's urban and transport history

Atlanta originated as the terminal point of the Western Railroad in 1839. After World War II, the city experienced an explosive growth and sprawled extensively. With the creation of a metropolitan planning agency in 1947 (later named the Atlanta Regional Commission (ARC)) regional planning and coordination among the intergovernmental agencies for the 10-county Atlanta region began (ARC 1996). According to Beaty (2007), however, despite the establishment of the ARC, there was rarely any visible evidence of coordination in Atlanta's subsequent development. The agency principally concentrated on the development of the central business district (CBD), which resulted in uncoordinated land use and weak regional facilities accompanied by a rapid outward expansion of the metropolitan area (Nelson 1993). Atlanta's history has been marked by a social and racial divide. By the 1980s, the inner city was dominated by an African-American population, most of whom were poor, jobless and homeless (Beaty 2007). Whites had moved to the suburbs, almost completely abandoning Atlanta's center. Redevelopment and revitalization of downtown Atlanta was on the political agenda, especially under mayors Andrew Young and Maynard Jackson in the 1970s and 1980s. As a solution, they sought to conquer the hospitality industry, starting with the Democratic National Convention in 1988.

Atlanta's transport system has been and still is mostly based on its extensive road network (Chapman 2000). Atlanta's freeway system was first planned in 1946 (the Lochner Plan) and continuously expanded, for example, through the National Defense Highway Act, which made Atlanta the juncture of three interstate highways (Helling 1997). Streetcars, trackless electric trolleys and automobiles were the primary means of transport until 1963, after which private cars and buses dominated the streets and finally displaced other means of public transport. Only in 1975 did the Metropolitan Atlanta Rapid Transit Authority (MARTA) start building a rail system, to open by the end of 1996 (Helling 1997: 75). Racial tension played a significant role in transit decision-making (Helling 1997: 74). In the *Economist*, a writer (Anonymous 1999) explained that counties populated with a majority of whites have ever since avoided anything that would connect and bring them close to the primarily black occupied CBD. In fact, this was "the real reason why links to MARTA (which was said in its early days to stand for 'Moving Africans Rapidly Through Atlanta') have been resisted and why, . . . people in the outlying suburbs stick doggedly to their cars."

Use of an automobile to access the suburbs was an unaffordable option for African Americans, forced to reside in downtown Atlanta. Any expansion of public transportation met resistance from political leaders, because with it would come "those people" (Beaty 2007: 41). With limited public-transit options available, the sprawling city had rapidly increased residents' dependence on the automobile, and investments in modern superhighways became a necessity for accessing Atlanta (Exhibition on Olympic Games, Kenan Research Center, Atlanta, USA). The highway system developed further, radiating from Atlanta's CBD in six directions (ARC 1996). "Traffic was nightmarish for commuters" (p. 19) and "transportation was legendarily chaotic, late, [and] crowded" (p. 31); thus Beaty (2007) describes the transport situation before the Olympic Games.

Given the socio-economic disparity in Atlanta and its impact on public and private transport services, Atlanta's future development plans only called for a small expansion of the MARTA's rail, but a significant upgrade in road infrastructure by 2010 (Map 4.1).

Atlanta's infrastructural plan clearly focused on improving highway infrastructure by building four-, six-, eight- and 10-lane highways. Until 2010, the city was planning to add 162 miles of new roads, to widen 759 miles of its

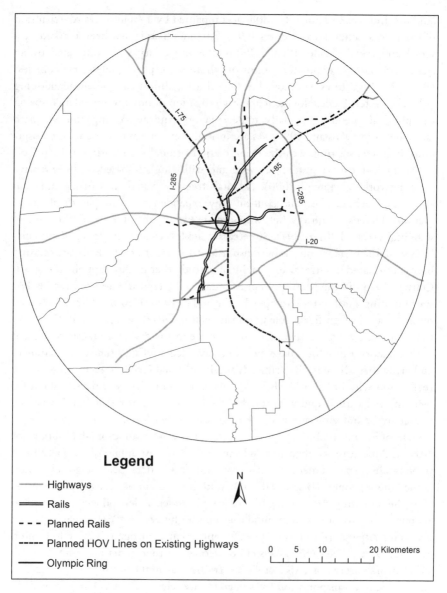

MAP 4.1 Atlanta's planned MARTA expansion, road expansion and HOV lanes

Created by author, overlaying multiple maps from ARC (Atlanta Regional Commission) 1999

highways, and to upgrade 87 miles of its road network. The plan also suggested 21 park-and-ride facilities along major highways and at MARTA's rail stations. Buses were to be the primary means of public transport, including radial and cross-feeder buses to MARTA. The buses and other vehicles shared the high-occupancy vehicle (HOV) lanes, which were to be implemented by 2010 according to ARC's master plan (see Map 4.1). Once Olympic planning began, further HOV lanes, on I-20 (the east–west connection) were added to the lanes of the original 2010 transport plan. The stimulus for these additional HOV lanes likely stemmed from the IOC's free flow traffic requirement for Olympic buses.

Atlanta's candidature and Olympic preparations

Aligned with the mayor's desire to foster Atlanta's hospitality industry, Billy Payne, a real estate lawyer in Atlanta, had the idea in the mid-1980s to host the 1996 Olympic Games. Embracing Payne's vision, he and his friends, informally known as the "Atlanta Nine," formed an informal bid committee that lacked any political support, because the local authorities did not believe Atlanta would stand a chance of winning the Olympic bid. In 1987, Atlanta submitted its bid (along with 14 other cities) to the United States Olympic Committee (USOC) to become the candidate city of the USA (Senn 1999). After winning the US bid, Atlanta entered the international competition for the Olympic Games in February 1990. In contrast to the public, private businesses strongly supported Atlanta's bid for the 1996 Olympics. Already in 1989, they began preparing for the possibility of Atlanta winning the Games. To take advantage of the potential monetary influx and the associated tourist spending, business owners fostered the creation of the Metropolitan Atlanta Olympic Games Authority (MAOGA), which was eventually established through House Bill 659 (Blair et al. 1996). MAOGA would become the public authority responsible for review of construction contracts, financial statements and budgets, approval of venue changes, and construction of the Olympic stadium. Public authorities, still, only reluctantly supported the bid, refusing to take on any financial responsibilities for the Games. In September 1990, the IOC awarded Atlanta the right to stage the 1996 Olympic Games (ACOG 1993c: 6–8).

The Atlanta Nine had been the driving force behind the bid as the core of the Atlanta Olympic Committee (AOC). In early 1991, the AOC reconstituted itself as the Atlanta Organizing Committee of the Olympic Games (ACOG), a private, non-profit corporation. Billy Payne headed ACOG, as president, with the former mayor of Atlanta, Andrew Young, as his Co-Chair (ACOG 1995a; Beaty 2007). ACOG was in charge of planning and staging the events for the 1996 Centennial Olympic Games. What distinguishes the Atlanta case from all other cities in this book is that Atlanta's government refused from the very beginning to financially participate in the Olympics. The city leaders' decision was driven by the reluctance of the public to back the bid. Atlanta's residents feared heavy tax increases due to the Games, because Montreal was still paying off its Olympic deficit from 1976. This significantly altered the planning approach for urban and transport legacies. Like Atlanta's US predecessor, Los Angeles (1984), the Games were to be privately financed (Preuss 2004). Therefore, the focus of the organizing committee was to raise funding and cut expenses to a minimum.

The CBD as the heart of Atlanta's Games

Political leaders saw Atlanta evolving through the Olympics into the next great international city (ACOG 1993a; French and Disher 1997; Yarbrough 2000). With this vision, it became necessary to revitalize Atlanta's CBD to display a clean and glorious image of a world-class city. With the Atlanta Nine as the growth regime in place, the Olympics "gave the business/political power elite the excuse they had longed for to tighten their grip on Atlanta's development" (Beaty 2007: 6). The result showed, as Short (2004: 107) argued, in a push for "major urban restructuring that primarily benefited business and corporate interests."

To rejuvenate and win back Atlanta's CBD as a thriving business center, Atlanta's urban plan for the 1996 Centennial Olympics had as its core idea to concentrate most of the Olympic venues in the center of the city. This Olympic Ring, which was an area conceived as within a 2.4km radius, contained 11 Olympic venues, in which 20 sports were to be held. The main Olympic Village, located at Georgia Tech, the Olympic Family hotel, and further housing for athletes, officials and the Olympic Family was spread within the Olympic Ring. In the same area, the Olympic media had its headquarters, with the main press center and the international broadcast center located in the Georgia World Congress center (Map 4.2).

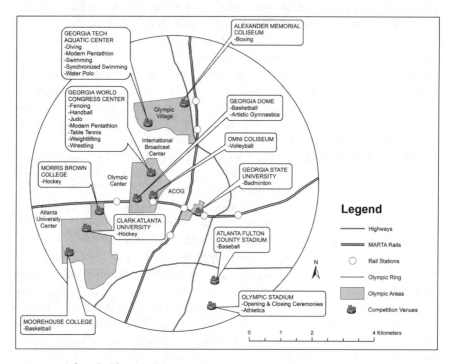

MAP 4.2 Atlanta's Olympic Ring

ACOG 1995b

Concentrating most Olympic activity in a relatively small ring seemingly offered easy and fast access to most venues for all client groups (ACOG 1995b: 7). Because the only two subway lines of MARTA intersected near the center of the Olympic Ring, spectators could walk to sporting competitions. Despite the advantages proclaimed by ACOG, the high concentration of 80 percent of Olympic activity in the center of the city eventually caused significant challenges in transporting athletes and officials to their respective venues during the Games (as we shall see later). Also included in the bid was the creation of Centennial Park, which was to become Atlanta's most visible imprint of an urban legacy. ACOG sought to create a park in downtown for the Olympic Games as "a gathering place for hundreds of thousands of visitors during the 1996 Olympics" (ACOG 1995a: 79).

With most of the venues being located in the central city, this circular set-up promised the revitalization of 15 decaying neighborhoods. In charge of planning and executing Atlanta's revitalization efforts in the downtown neighborhoods was the Corporation for Olympic Development in Atlanta (CODA) (Ruthheiser 2000). Their efforts, however, primarily focused on creating a worthy Olympic image. Without public funding at hand, AOC built new homes only for the most visible, front-row communities, becoming "living advertisements for how well Atlanta treated its poor neighborhoods" (Beaty 2007: 6).

Atlanta's transport preparations

"Atlanta sold its bid principally on its transport" believes Powell (23 October 2007), the athletics correspondent for *The New York Times* for 18 years. The IOC had doubted that Atlanta's strongest competitor for the Centennial Games, Athens, would be able to handle the traffic volume associated with the Games, given the infrastructure at the time of application (Senn 1999). The expectation for a perfect transport performance was set high from the very beginning. ACOG promised that transport would "allow the guests to move comfortably from event to event breaking new ground in efficiency and techniques and surpassing past Olympic transportation systems in efficiency" (ACOG 1993b: I-1).

The Olympic legacy for transport was to "establish an integrated, multimodal transportation system which moves people and goods in a more efficient and environmentally sensitive manner" (CODA 1993: 3). Organizers foresaw as the long-term benefits to the city "repairs to key transportation facilities, an advanced traffic management system, high occupancy vehicle lanes on the freeway system, the multimodal Atlanta transportation centre, travel behavior changes, pedestrian facilities, the use of alternative fuel vehicles and a continued spirit of cooperation and pride" (ACOG 1993b: I-7). Advertised to the IOC throughout the bidding, candidacy and preparation period was the technology and advanced infrastructure with which Atlanta would present itself to the world (ACOG 1995b). Specifically, ISTEA, the Intermodal Surface

Transportation Efficiency Act of 1990, allotted additional funds for the Atlanta region to develop the Intelligent Vehicle Highway System (IVHS) demonstration project. With these improvements in place the city was expected to run "like clockwork, ensuring mobility for all" during the Olympics (ACOG 1993b: I-1). In the long term, ACOG and residents hoped that the transport changes would result in different commuting patterns, thereby becoming "the legacy of the Games for transportation" (Kelly Love cited in Goldberg 1996).

Transportation planning for the Atlanta Games began in 1989, when a group of transport agencies developed the first 1996 Atlanta Olympic Transportation plan (ARC 1996): at the time, Olympic venue locations had not been fixed. In 1992, ACOG, which was responsible for providing transport to the Olympic Family and spectators attending the 1996 Centennial Olympic Games, officially put two agencies in charge of transport planning: the temporary Olympic Transportation Support Group (OTSG) and the well-established Atlanta Regional Commission (ACOG 1993c). Throughout the candidacy and even shortly before the Opening Ceremony, important locations continued changing. Without any planning security for training facilities, parking lots or bus depots, robust and reliable transport planning was nearly impossible (ORTA 1998: Appendix A, ii). Furthermore, traffic estimates kept rising as Atlanta approached the Olympic Opening Ceremony. In the end, planners grossly underestimated the actual demand for Olympic transport services, which eventually led to traffic congestion and severe overloads on transit systems during Games' time. In retrospect, Atlanta had staged the largest and most compact Olympic Games ever in Olympic history (Dahlberg 1996).

Once ARC and OTSG completed planning, ACOG had to coordinate and manage an integrated Olympic Transport System (OTS). The committee chose to do so via three different transport networks (ACOG 1994; U.S. Department of Transportation 1997):

- The Olympic Family Fleet System (OFFS) consisted of the Olympic Family motor pool, primarily automobiles and other light vehicles.
- The Olympic Family Transport System (OFTS) provided transport for 150,000 people, including all athletes, technical officials, NOC and IOC members (ACOG 1995a: 11). ACOG outsourced this service to a private bus company that had no experience in handling large passenger volumes.
- The Olympic Spectator Transportation System (OSTS) consisted of the 46-mile MARTA rail system with 36 stations and MARTA's 150 bus routes. Under contract to ACOG, MARTA operated the OSTS – defined as the interconnected network of park-and-ride lots, shuttle buses and rapid rail cars that would serve all of Atlanta's venues (Bulkin 1995: 25). On this network, express buses operated from park-and-ride lots to the Olympic Ring, and shuttle buses ran from selected rail stations to outlying venues.

Shuttle buses also served outlying venues 24 hours a day from selected park-and-ride lots (MARTA 1996).

All three transport systems were run on a one-way street system in downtown Atlanta and on the high occupancy vehicle (HOV) lane system on the highways (ACOG 1993b; U.S. Department of Transportation 1997). For spectators, transportation to and from the venues was free with an Olympic ticket (Dahlberg 1996).

As in Barcelona's case, the Olympics required infrastructural investments in Atlanta's transport systems (ACOG 1993b: III-1–III-4). In the run-up to the Olympics, Atlanta constructed freeways and arterials at a faster pace than ever before, while MARTA completed more than 20 major projects in time for the Olympic Games (Table 4.1).

The Olympic momentum had tremendous impact on the pace with which road projects could be implemented. According to Waters (Interview 2009), a state traffic engineer for the Georgia department of transportation prior to, during and after the Olympics, it "allowed the city of Atlanta and the department

TABLE 4.1 Atlanta's transport investments for the Olympics

Road infrastructure	Public transport	ITS
Ten road construction projects in the inner city	Extension of the East Line services through Kensington to Indian Creek Station in June 1993	Advanced traffic management system, including combined state-of-the-art traffic control, freeway surveillance and an incident management system
Street improvements in downtown	Three new stations: Buckhead, Medical Center and Dunwoody	
High occupancy vehicle (HOV) lanes*		IVHS to accommodate the increase in the expected traffic flow but also to represent Atlanta as a clean new city to the visitors
Olympic landscaping† of many sections of the I-20, I-75 and I-85	Enhancements at transit rail stations	
Renovation and enhancements of key bridges	MARTA mid-life overhaul, including an institution of automatic train announcements	multi-modal transportation center
	Perry Boulevard CNG Bus Facility	

Sources: Compiled by author. Adapted from MARTA 2009; ACOG 1994.

Notes:

* HOV lines are dedicated to vehicles with two or more persons, emergency vehicles, motorcycles and buses. The first HOV lanes opened in mid-December 1994 (18 lane miles on I-20 from downtown to I-285). In 1996, 60 additional lane miles opened on I-75 and I-85 (Department of Public Safety 2009).

†Beautification along arterial and inner city roads.

of transportation to advance its ITS plans and to implement them not over a 10-year period but over a four-year period." Besides this catalytic affect, Waters acknowledged a change in the original road plans:

> the implementation of the HOV lanes on I-75 and 85. HOV lanes had been planned for many years; [they were] just redone in the fact that [planners] utilized an existing lane and its shoulder width to add another lane rather than take away a lane that had been designated before.
>
> *(Interview Waters 2009)*

Besides upgrades to road systems, other transport investments in public, active and air transport infrastructure prepared the city to host the Centennial Games. For the first time, the MARTA rail line went beyond the I-285 perimeter highway to enhance access to Olympic venues at Stone Mountain, the host of Georgia's international horse park (Map 4.1). In downtown, 12 projects for pedestrians were completed in record time. In addition, new bike paths stretching 17 miles connected the Olympic Village at Georgia Tech to Stone Mountain Park, where a smaller cluster of venues was located. Hartsfield Atlanta International Airport also gained $500 million of improvements, including a $350 million international air terminal, a $24 million atrium and a new extension of the underground train to connect with Hartsfield's concourses. All the projects were part of the airport's master plan, but the Olympics created a catalyst to move the projects forward sooner (Dahlberg 1996; Vaeth 1998).

Despite these investments, operating the Games, as the organizers were very well aware, posed the risk of heavy congestion in the inner city due to Atlanta's large share of private transport. Creative strategies to reduce demand during the Games were essential, based on the understanding that "a significant reduction in normal daily traffic into and through the city [was] necessary in order to make room for Olympic-related traffic and prevent gridlock" (Dahlberg 1996: 11). The key characteristic of Atlanta's transport system during the Games was a heavy reliance on pedestrian movements and public transportation (ACOG 1995a: 80; ORTA 1998). Furthermore, during the Games the city promoted biking and walking as primary travel modes to Olympic Ring venues. Knowing that Atlanta had a car-dependent population, ACOG ran a systematic communication campaign to shift residents to public transit. The committee also urged employers to institute flexible working hours, ride-share programs, offer tele-commutes and other efforts to help control traffic, basically minimizing the number of trips wherever and whenever possible.

Transport performance during the 1996 Olympics

Atlanta hosted the Centennial Games from 19 July to 4 August 1996. Atlanta became known as "the chaos Games" due to significant glitches in the transport and technology systems. "If Atlanta was not the best, it may rank in one of the top three Games, where the most world records were achieved. However, Atlanta is mentioned not for its tremendous successes, but because of the transport failure," said Protopsaltis, the head of transport operations during the 2004 Olympics (Interview 2007). The poor transport performance had led to athletes missing competitions, angry spectators waiting to get on MARTA, and endless drives on the gridlocked freeway system (Applebome 1996; Rushin 1996). In the closing ceremony, Antonio Samaranch, the IOC president at the time, said Atlanta had hosted "great Games," but he stopped short of saying they were "the best Games ever," the usual phrase with which the president ends the Olympic Games.

"Most of [the transport problems] were Olympic Family members and media access problems," according to Graham Currie's (Interview 2008) assessment of the transport problems in Atlanta. Because journalists had to travel between venues mostly by buses and shuttles with ordinary spectators, the long waiting hours in the humidity and heat of the city and the 15-minute walk to the MPC explain "the headlines of some newspapers on the 'great Olympic chaos in Atlanta' " (Billouin 1997: 161). Another factor, or so Currie (Interview 2008) argues, is that the media always needs a great story (the need for such a story was confirmed by several other interviewees wishing not to be cited).

> One of the facts is that the international media pays for the Games. A simple fact is that they will find a story. They have to find one. The story they found was, before the Games, many people from different organizations saying how good the Games were going to be, [and] then, during Games time, a city struggling with over demand.
>
> *(Interview Currie 2008)*

The problems with media and athlete mobility emerged early on in the Games. ACOG "had run the Olympic Family transport system so poorly that the IOC directed organizers to fix it, and fix it fast" (Hill 23 July 1996). The negative press and the consequent damage to Atlanta's image led Olympic advisors to demand that the Olympic Family and particularly the media are well catered to in future Games (Currie 2007). The IOC from thereon out set new requirements for more buses, newer buses and excellent road conditions during the Olympics that guaranteed unhindered access to each of the venues by private vehicles.

Spectator transport also experienced major glitches. At the beginning of the Games, people complied with the imposed temporary traffic laws (Hill 23 July 1996). After a few days, the situation started to deteriorate, presumably because commuters got wind of "gypsy parking" spots in the inner city and opted for their cars instead of the overcrowded public transit systems (Hill 29 July 1996). More evidence for the change in Olympic commuting behavior came from journalists observing "vacant parking spaces at MARTA lots" (Goldberg and

Hill 1996). In short, traffic congestion occurred from people driving despite the urgent message not to: the advertising campaigns for public transit failed miserably (Monroe 1996). Interestingly, in ARC's post-Olympic report, a quite different transport picture of Olympic transport performance is painted:"traffic congestion was almost non-existent except for a few recurrent hot spots" (ARC 1996: 37). Overall, three shortcomings led to the poor performance of Atlanta's transport system: the underestimation of spectator travel demand on MARTA, the undersupply of buses and the uncoordinated transport modes during the Games.

First, transit estimates were well below the actual attendance. MARTA carried 15 million passengers instead of the expected 9.6 million (Booz•Allen & Hamilton 1997). This "increase of close to 50 percent beyond expectations," in Currie's (1998: 7) judgment, "proved impossible [to handle] without significant service quality deterioration." As an example of the task at hand: MARTA's usual weekday load was 230,000 boardings, the expected Olympic average was 588,000, the actual Olympic average was 953,000 boardings, four times the normal weekday load (Booz•Allen & Hamilton 1997). The maximum load MARTA carried on its peak day came close to 1.3 million passenger trips per day (Monroe 1996). In Currie's opinion (Interview 2008) MARTA "did amazing things during the Olympics. It carried volumes you would never ever have thought possible, but it was far too small to carry the [Olympic] volumes."

Second, besides MARTA, the Olympic transport system heavily relied on buses. Initially, Atlanta had asked for 2,000 new buses, but only received 1,800. The buses, which transport organizers had to borrow from other cities and states, were expected to be in excellent condition worthy of transporting Olympic family members. But the actual vehicles they received were old and experienced frequent break downs (Billouin 1997; Booz•Allen & Hamilton 1997; Monroe 1996). To run the Olympic bus systems, over 2,000 additional bus drivers were required. These could only be obtained from sources "out of state," unfamiliar with the road network in Atlanta (Billouin 1997; Booz•Allen & Hamilton 1997). The media complained about unfriendly bus drivers who abandoned their buses on the road, describing the phenomenon as inexplicable and rude. Currie (Interview 2008) offered a different part of the explanation: "when you have under manning of these resources you end up having the drivers [having] to work harder. As a result, they are getting very tired. You have reports from the media that they are unreliable, not turning up, and lazy. It is not because of that, they are simply exhausted."

Third, the Olympic transport systems were managed and organized by three transport entities (Gonzalez et al. 2000). This led to miscommunication and unresolved responsibilities during Olympic operations. As a result, Booz•Allen & Hamilton (2001: 43) recommended in their advisory report for Sydney based on the Atlanta Olympics, that "special event transit operations should be managed under a single organizational umbrella."

Transport outcomes and Atlanta's legacy

Atlanta is left with a bitter after-thought to its poor transport performance. For Currie (Interview 2008), the facts of why Atlanta experienced such transport problems were clear:

> If you look at the simple fact of Atlanta as a city: it is not a good Games city, it has an extremely poor public transport system and very poor bus system, and there is another fact in that the Atlanta Olympic Games was one of the largest ever. What we got here is a city that is not designed to handle mass volumes at all, handling the biggest volume ever.
>
> *(Interview Currie 2008)*

Yet understanding how this apparently obvious fact could have been overlooked – by the IOC and Atlanta – is worth revisiting. The IOC took the stance that Atlanta's organizers had not listened to the advice the IOC had given the city during Olympic preparations. "The IOC was putting the responsibility on the ACOG, with some IOC members saying that the Americans had refused to listen to advice or pay any attention to previous experience" (Billouin 1997: 161). In interviews conducted after the Games, Atlanta organizers also blamed themselves: "we have never seen a single document from Barcelona, because we were too lazy to translate" (Interview with Macey 2003). So, this is the official IOC story.

Archival documents from the IOC and interviews, however, shed a different light on the story. In fact, the IOC never saw a problem in Atlanta's transport preparations. The evaluation commission of the Association of Summer Olympic International Federations (ASOIF) visited Atlanta in 1990 and concluded that "no major traffic problems can be expected in Atlanta for the '96 Olympics" (Boitelle et al. 1990: 12). Furthermore, ASOIF did not detect any major issues or potential pitfalls with public transport operations during the Games. The impression the public transport system made on the IOC evaluation commission was also very positive: the traffic conditions were experienced as fluid and MARTA as a rapid, efficient modern transport system (Ericsson et al. 1990). In fact, MARTA was expected to carry high Olympic peak loads easily (ACOG 1992). In contrast to Macey's (2003) claim, Atlanta planners did observe each organizational effort of the Barcelona Games, with over 100 delegates (ACOG 1993a, 1993b). Vilalante (Interview 2007) confirmed ACOG's reports in Barcelona; he remembered that the commissioner of the city of Atlanta was his co-pilot during the transport operations of the 1992 Games.

This unofficial story sheds light on the rather minimal influence the IOC had on Atlanta. Because the IOC did not anticipate any transport problems, they did not see a need to interfere in the transport planning process prior to the Games. Hence, most of the projects that had been implemented into Atlanta's transport system during the Games had a long-term perspective:

All of the projects, [except] one,* that I can think of were sustainable and were in the overall transportation projects list. The only exception was a different exit into the Stone Mountain Park on US 78. . . . And the state did build that. The anticipation was that it would become an emergency or overflow exit.

(Interview Waters 2009)

Furthermore, the ITS technologies, which were a "shining star" for the bid and during the Olympics, have remained as a key transport legacy for Atlanta and the entire region, stated Dunn (Interview 2009), the ARC's spokesperson for external relations in the transportation planning division. But she admitted that "with funding challenges, keeping it up to speed has been an issue." From a transport perspective, Waters (2009) could not recall any instances in which Atlanta could have gotten more out of the Games than it eventually did. In retrospect, he described the planning approach as follows:

The professionals that have been involved in this have asked themselves over and over and over again [whether they could have done anything different]. The answer is "No." Most of the professionals in the Atlanta area became involved very early on and mutually engaged in a pledge that we would do everything that we could, humanly possible, to make the Games a success. We did not want to have the situation after the Olympic Games, where we would say "if we had just tried a little harder, we could have done better." I sincerely believe we did the very best we could and we fully utilized every resource that was made available to us.

(Interview Waters 2009)

From today's perspective, there were not many opportunities Atlanta had missed given the public reluctance to back the bid. Even though revolutionary changes to Atlanta's transport system could not be undertaken as those require substantial public investments, the city did progress significantly with its own urban master plans. In the run-up to the Games, private, public, active and air transport infrastructure received substantial investments that remained as lasting transport legacies in Atlanta's vicinity. However, the primary focus on road transport and negligence of the city's multimodality goal, led and will lead Atlanta's future development. The city's urban legacy was the rejuvenation of its downtown area, with the Centennial Park as the new public space. Office buildings and businesses eventually followed and turned the previously degraded area into a livable and striving center.

---

* This one exception that was, according to Waters, implemented solely for the Olympics did indeed become relatively useless for the city after the Games: "[The exit ramp] was added because of the Olympics and then it was abandoned about five years after the Olympics."

Conclusions

Because Atlanta's government shied away from backing the bid, it missed the opportunity of the Olympics to bring about major change in how people would travel in Atlanta in the future. Investments went primarily into expanding the existing road network and a new traffic management system. Public transport companies catalyzed only minor extension of existing lines through the Games, which did not deviate from their original plans. Dunn (Interview 2009) and Waters (Interview 2009) agree that the Olympics overall expedited transport projects that had been planned before the Games, and that Atlanta used the resources it had wisely for the future of its transport system beyond the imme- diate short-term needs of the Olympics.

A smart choice Atlanta had made to accommodate Olympic travel flows was overlaying the expected Olympic peak passenger flows onto the regular commuter flows. This required minimal investments in new public transport infrastructure and likely resulted in sufficient transport systems capacity – so both the IOC's evaluation committee and the Atlanta Organizing Committee (AOC) believed. Hence, the changes the city had to undergo solely because of the Olympic requirements deviated from the original city plans only to a minimum. In retro- spect, the investment choices in transport were beneficial to the city, but came at a high price for Atlanta. Not only did Atlanta's residents remain car-dependent, but also the city was known across the world for its poor Olympic transport performance. Currie comments on the choices that Atlanta made:

> What Atlanta did was to invest into a railway extension, which was of modest benefit to the Games. And they invested a lot in technology, such as the new traffic center. What did that do for the Games? You could not get the volumes of people in on the road system; you needed a transit system. Another example of this is that Atlanta implemented a lot of resources on smart cards/ticketing. Of what benefit is that? What it did was take a lot of resources away from the Games.
>
> *(Interview Currie 2008)*

The IOC learned three important lessons from Atlanta's transport fiasco: first, media had to be handled with kid gloves, since the press was crucial for broad- casting a positive image to the world. Therefore, in future Olympic Games, the press received dedicated buses and cars running on exclusive bus lanes. Second, extensive mass transit systems are necessary to cope with the peak passenger demands for the Olympics. This lesson, as we shall see in the latter case studies, had a tremendous impact on existing transportation plans. Third, a single entity for Olympic transport was essential, with responsibility for overseeing transport operations and equipped with cross-institutional power.

5

PLANNING FOR THE 2000 OLYMPICS AND SYDNEY'S URBAN LEGACY

Sydney barely used the Games to foster its transport vision and failed to implement the holistic and comprehensive urban development strategy that the city's planning department had laid out in its master plan. What Cashman (2006) teasingly titled Sydney's "bitter-sweet awakening" also holds true for the transport legacies of the millennial host city Games. In the run-up to the Olympics, Sydney invested in urban freeways and rail systems that accommodate Sydney's development strategy only to a very limited extent. As the center of Olympic activity, Sydney chose to transform a heavily polluted brownfield into a greenfield, Homebush. While this project was on the drawing board for years, the catalytic Olympic momentum that truly could have transformed Homebush, the center of Olympic activity, was not used to its optimum shortly after the Games. In particular for transport, planners took a short-sighted view. To make Homebush accessible for Olympic peak loads, a rail loop, stations and access roads had to be built. These investments, however, did not integrate the area into the regional structure of Sydney's northern suburb, but rather were an add-on to the transport system designed by Olympic transport needs. Since then, Sydney has made multiple efforts to transform the Olympic infrastructure into a viable and thriving legacy for the future. Since the Games, Sydney has made strides to improve its legacy and successfully transformed Homebush from a polluted brownfield into beautiful parklands with lively communities (Cashman 2011).

Sydney's urban and transport history

Starting in the early 1950s, Sydney and its greater metropolitan region experienced unprecedented growth stimulated by the postwar economic boom. Industrial clusters spread along the coastal areas; commerce and retailing were located in central Sydney with coastal sub-centers in Newcastle, to its north, and in Wollongong, to its south. Public transport was the primary means of travel anchored in Sydney's central business district (CBD) at Central Station, from which a radial public transport system served surrounding areas and outer neighborhoods (Punter 2005). The rapid growth of the region changed Sydney's character as it continued to grow substantially through the 1960s, 1970s and 1980s. High-rise commercial buildings flourished in the CBD, while people moved to the suburbs, driven by the desire for detached homes. With this population shift came the rise of the private vehicle. Eventually, the tramway lost its competition with the private car. Physically aged, run-down, and judged to be unsuitable for their radial role in Sydney, the tracks were removed and replaced by buses: essentially buses could perform the same service as trams but were more comfortable, cheaper, safer, faster and more efficient (Gibbons 1983). In the past 20 years, this trend to population sprawl has followed the relocation of employment activities, which led to an inner-city decline of manufacturing, making room for the new service sector. As Sydney continued to grow beyond its original boundaries, the patronage of the rail systems declined further, resulting in a 12.8 percent drop in the number of trips taken by public transport between 1981 and 1991 (Gee et al. 1996: 14). Throughout this transition, Sydney has kept its old CBD, and even though the city's residents remain largely car-dependent, high-density development has occurred along existing rail corridors (Department of Planning 1995).

Planning played a central role in Sydney's development. Following the County of Cumberland Plan (1948), Sydney's first attempt at land development of the region, the Sydney Region Outline Plan of 1968 proposed two transport strategies to guide the future growth of the largest metropolitan area in Australia. First, it was proposed that linear expansions along communication and rail corridors should be fostered. Second, a strong Sydney–Newcastle–Wollongong corridor along the coast was envisioned as one interrelated linear urban complex (Department of Planning 1968). In transport, meeting current and future car demand was a priority (Department of Planning 1988, 1995). Transport and its integration with land-use along with an emphasis on environmental protection became one of the top priorities for Sydney in the late 1980s; because the lack of such integration had reinforced Sydney's outward growth, the region's road infrastructure demanded continuing investments and increased maintenance (Department of Planning 1992). Therefore, extending the public transport system and its services, while simultaneously reducing the need for new roads, were new key targets for urban planning (Department of Transport 1995). As Sydney spread upstream towards its hinterlands, secondary employment centers such as Parramatta had started to evolve (Map 5.1). Pursuant to a Department of Planning discussion paper (1993), a new emphasis was placed on strengthening Parramatta's role as a secondary employment center close to Sydney. Parramatta is located 23km west of Sydney and close to extensive parklands and the wetlands of Millennial Park. Sydney's metropolitan expansion strategy had as its main goal to create one region integrating several of the Sydney's satellite cities via a regional transportation network.

Sydney's planning and transport departments jointly identified strategic transport opportunities in the early 1990s (Map 5.1). From a transport perspective, Sydney and

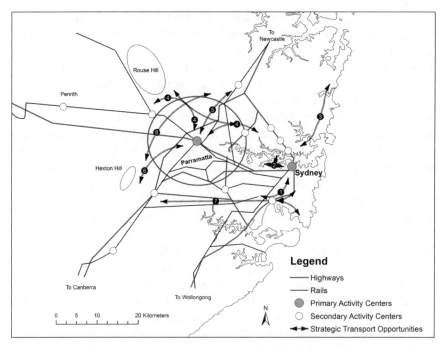

MAP 5.1 Sydney's strategic transport opportunities

Created by author with GIS (country boundary layer) and georeferencing points according to multiple Google maps. Original source: Department of Planning 1995: 99 © State of New South Wales through the Department of Planning

Parramatta presented the best opportunities to encourage public transport use (Department of Planning 1995; Department of Transport 1995). Parramatta being the second largest employment center in the Sydney region, the Department of Transport suggested making the city a major bus–rail interchange and establishing new transport corridors from the lower North Shore, Hornsby, Rouse Hill and Hoxton Park (Department of Transport 1995: 36–41). Throughout the following years, other departmental plans called for the promotion of public transit and reduction of individual reliance on cars (Department of Urban Affairs and Planning 1997).

Sydney's candidature and Olympic preparations

Sydney's aspiration to host the 2000 Games started in mid-November of 1990 when the New South Wales (NSW) government informed the head of the Australian Olympic Committee that Sydney was interested in bidding for the 2000 Olympics, pending the Baird report, which assessed the financial feasibility of Sydney staging the Games. This report came out in December of the very same year, concluding that the Games could be staged at a profit. That same month, the federal government announced it would release 84 hectares of land at Homebush Bay for the Olympics, which was to be the prime location for Olympic facilities (Whitten 2000).

Other location choices followed, building a linear pattern along an east–west axis stretching from Penrith in the west, to Holsworthy, Bankstown, Bondi and Darling Harbour in the east (Map 5.2). Most facilities were located in two primary Olympic Zones (14km apart from each other), Homebush and Sydney Harbour. Twenty-one out of 25 competitions would take place in the two centers, with 80 percent of all the events held in Homebush (De Frantz et al. 1998).

The IOC transportation working group assessed that these location choices would create heavy travel demand between the Olympic Park at Homebush and Sydney's CBD (De Frantz et al. 1998). At the time, the Sydney Olympic Park area and the Sydney Harbour zone were located relatively close to road systems, rail networks and water-based services. However, given that Sydney residents were generally car-oriented, the IOC working group believed that significant communication efforts would be required to divert drivers towards transit systems. The inquiry commission of the International Olympic Committee (IOC) praised Sydney's bid, because it presented well-thought-out planning and required a maximum travel time of 30 minutes to all venues (Ericsson et al. ca. 1993: 63–5). The commission also noted that the bid offered conditions beyond what was actually required by the IOC.

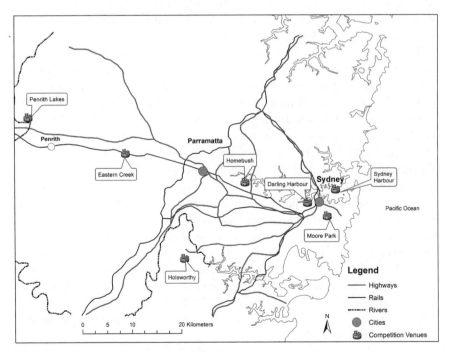

MAP 5.2 Sydney's Olympic venue locations

Created by author with GIS (country boundary layer) and georeferencing points according to multiple Google maps. Original source: Department of Planning 1995: 36 © State of New South Wales through the Department of Planning

The Department of Transport (1995: 54) assured the general public that "Olympic transport plans will . . . be consistent with integrated plans for the region explored in this [transport] strategy." The Sydney Organizing Committee for the Olympic Games (SOCOG) reiterated this goal in the strategic Olympic master plan, and the Olympic Coordination Authority (OCA) continued to vow in their state-of-play report to the people of NSW, "to make the most of the opportunities the Games present to the city" (OCA 1997: 9). In contrast to these promises, Sydney's original metropolitan plans laid out different urban and transport opportunities that did not match well with the Olympic travel forecasts. Because the Olympic Games fostered an east–west connection on existing rail lines, Sydney could not realize most of the strategic transport opportunities (primarily new north–south connections) identified by the Department of Transport (Map 5.1) in the run-up to the Olympic Games.

Persons put in charge of running transport, such as the director for strategic planning within the Olympic Road and Transport Authority (ORTA), Steve McIntyre (Interview 2008), recall a quite different approach than OCA's promises to the public in regards to the transport-planning process. Given that ORTA was created late in the preparation for transport for the Games, he argued he had to take a very pragmatic approach. "Legacy issues for Sydney were second-order; the first-order issues were that we actually understood the scope of the task, that we defined responsibilities for what needed to be done, and that we had a robust planning approach" (McIntyre Interview 2008).

In the run-up to the Olympics, Sydney's planning institutions lost most of their executive power. Due to the streamlining of planning controls for the Olympics, the level of control held by local councils over their respective jurisdictions had been significantly reduced. Throughout the preparations for the Olympic Games, this caused some conflicts between local councils and the state government, which was "single-minded in its approach to Games planning" (Connor et al. 2000: 9). For the Olympics, decision power laid within the Sydney Organizing Committee for the Olympic Games (SOCOG), which was also the communication center between the IOC, Australia's Olympic Committee (AOC) and the governments (Dunn and McGurick 1999). SOCOG's decision-making power trickled down to agencies with specific functions, which were legislatively empowered to direct state and local agencies. In charge of running the Games on the ground was the Olympic Coordination Authority (OCA) established in 1995. OCA was a cross-department body set up to ensure a coherent approach to the Olympic planning process, including the task to identify and realize a legacy for Sydney and NSW. Equipped with authoritative control, OCA was in charge of master planning and developing the Games' sites (Dunn and McGurick 1999).

As a result of the Atlanta traffic chaos the IOC Transport working group pushed for the creation of a centralized Olympic Road and Transport Authority, in short ORTA (De Frantz et al. 1997; SOCOG 2001). Shortly following ORTA's creation, on 7 April 1997 a fundamental restructuring of transport planning and operational structures took place in the spirit of the Sydney Olympic Games. ORTA's task was to integrate and coordinate transport to and from all competition and non-competition venues and the city centre (Curnow 2000; ORTA 1998b). Existing transport agencies such as the Roads and Traffic Authority (RTA), State Rail Authority (SRA) and State Transit Authority (STA), along with the private service providers Bus 2000 and Coach 2000, had to operate transit according to ORTA's plans and specifications (ORTA 1999b). In principle, ORTA had two clients: the SOCOG for Olympic Family transport, and the NSW Government for transportation of spectators, visitors and the general public commuting in Sydney. ORTA also believed that the Olympics provided Sydney with a unique opportunity to promote maximum public transport use and launch innovative transport management practices (James and Myer 1997: 9). ORTA's hope resided in the assumption that if the transport system performed well, more people would use public transport regularly even after the Games.

Sydney marketed itself to the Olympic Committee by placing a strong emphasis on environmental protection and promising to stage "Green Games" (ORTA 2001). Urban and transport planning, just like all other Olympic changes, were to be guided by strict environmental guidelines through the State's Environment Planning Policy No. 38 (Dunn and McGurick 1999). A year prior to the 2000 Games selection process (in 1992), the IOC had adopted the principles of sustainable development set forth at the UN Conference on Environment and Development in Rio de Janeiro into their Charter (IOC 1999: 9). Retrospectively, Sydney's choice of the "green" theme may have increased Sydney's chances of winning the bid.

Homebush Bay as the center of Sydney's Olympic Games

One of the key "green" theme selling points was the selection of Homebush as the heart of Sydney's Games. In retrospect, Homebush was to become Sydney's prime urban legacy. Heavy industry had abandoned Homebush by 1988, leaving the area polluted, degraded and contaminated – a brownfield. Homebush was flood-prone and hence unattractive to private investors, yet its surroundings were beautiful. The NSW government had been wanting to purpose-clear Homebush and had been struggling since the mid-1970s to choose between three general development ideas:

1. The Bunning Scheme (1973) proposed Homebush as a host for international sporting events. Among the other 19 alternatives around Sydney, Bunning had identified Homebush as owning significant advantages: its size, its centrality to Sydney, its nodal location in the transport network, its public ownership of the land and the relative lack of development (Waghorn 2000; Weirick 1999). Starting with Sydney's intention to bid for the Olympics, more and more studies supported this scheme, including the

Premier's department scheme (1988) and the Homebush Bay development strategy (1989).

2. The MacLachlan scheme (1982) combined three development purposes at Homebush – employment, residential and recreational (Waghorn 2000).
3. The Hub Scheme (1983) proposed that Homebush should be Sydney's first technology, industry and business park with industry and employment zones. Just before bidding for the Games, Sydney had bid for the "multi-function polis" (in 1987), an Australian–Japanese cooperation. Under this scheme, Homebush was to become a center for future-oriented high-technology and leisure facilities (Black 1999; Searle 2008; Waghorn 2000).

Due to the Olympics, the Bunning Scheme prevailed and already in 1992, a year before the IOC had to select the 2000 host city, work in Homebush Bay had begun. It was planned to contain 14 of the 28 sports venues, all of which had to be constructed in record time. In close proximity to Homebush, the Olympic Village was designed in the Newington urban district, housing all athletes and team officials, some of the judges and the media.

Plans for Homebush's legacy officially existed: the venues were built for multipurpose usage after the Games and Homebush's master plan (1990) suggested the relocation of the Royal Easter Show to Homebush. At the time, this yearly agricultural show drew hundreds of thousands of visitors to Sydney' Moore Park in downtown. Because Homebush was designed for the throughput of large numbers of people, later usage of facilities could also include exhibitions and rock concerts (Department of Transport 1995: 54). In the broader scheme of metropolitan planning, Homebush was to mirror Moore Park close to the Sydney CBD. As Moore Park was to Sydney, Homebush was to be to Parramatta (Waghorn 2000). Early on, these legacy prospects were questioned given the concurrent planning approach. Myer (1996), for example, could not identify the long-term planning approach the authorities had pledged to take: "the primary motivation for most of the current activity at Homebush Bay is meeting the requirements of the 2000 Olympics – and longer term planning issues generally appear to receive a lower priority" (Myer 1996: 2). The following sections confirm that Myer's impression of the legacy-planning process pre–Olympics became reality for most of the changes Sydney undertook. However, once the Games were over, Sydney fostered multiple initiatives to advance living conditions in and around the park and developed a comprehensive urban development theme post–Games (Cashman 2011).

Along with planning Homebush came suggestions on how to provide transport services. According to Juliet Grant (Interview 2008), a transport strategist for Homebush, the key concerns were the green theme, the capacity needed for the Olympics and a long-term vision for the site. After considering a variety of transport alternatives including exclusive bus transport via an extensive road system, light-rail access and heavy rail lines, the cabinet decided on the presumably optimal solution (Interview Grant 2008): a rail loop into and rail station for

Homebush, a ferry wharf and an extensive road network for access to and mobility within the Olympic park accompanied by many cycling and walking paths. In the end, to provide Homebush with a transport network suitable for the Olympics, (AUD)$97.7 million for rail and (AUD)$190 million in road would have to be invested (Jacana Consulting Pty Ltd 1996). As with planning for Homebush's future, the same short-sightedness held true for transport. As Daniels (2008) confirmed in an interview:

> The focus becomes so much on the event itself, the completely immovable deadline, and so you have to have everything done by then. Most people focused on what we have to deliver on this date. Not many people thought about what to do afterwards.
>
> *(Interview Daniels 2008)*

Within the few years of Olympic preparations, a massive new Homebush rail station and a rail loop for exclusive access to the sporting facilities, a new ferry wharf for athlete transport, and a road network to satisfy Olympic bus demands were built. All of these investments were not slated prior to winning the Olympic bid, but became permanent legacies that altered pre-existing transport plans. In retrospect, the Games not only had a purely catalytic effect on transport plans, but also altered them in favor of Olympic objectives. Many grassroots groups opposed the access alternatives that had been selected for Homebush. Greenpeace, the Clean Air 2000 Project team, and local communities argued for a proposal that built more public transit for the long-term, avoided further road construction, and created a public transit-oriented Olympic Village, all of which would reduce the need for local car traffic later on (Clean Air 2000 Project Team 1996; Greenpeace 1995; James and Myer 1997). Reports, objections and protests, however, came too late to have any impact on an already-made decision. The NSW minister for urban affairs and planning had given consent to the new rail and road infrastructure in February 1996, a year before ORTA was even founded. Given the time pressure to stage successful Games, there was not enough time to rethink the strategy and include a long-term site vision based on holistic strategic planning. This opportunity had passed; it could (and should) have happened either between 1995 and 1997 or during the bidding stage (Interview McIntyre 2008). And so, the following short-term solutions were implemented to accommodate peak passenger needs focusing on the Olympics and the Royal Easter Show.

In preparation for the Games, Sydney ultimately created multiple transport legacies. In 1995, OCA proposed a single rail loop of 5.3km in length, broadly following an old rail route (OCA 1995b). Searle (Interview 2008) recalls that in the beginning, Sydney's transport planning committee for Olympic transport had thought to handle the Olympic masses exclusively with buses. Likely because of Atlanta's poor transport performance, this idea did not resound well with the IOC evaluation committee (De Frantz et al. 1998). Therefore, construction of a rail loop connecting Homebush with the Main Western Line at Flemington Junction began in August 1996 (OCA 1995b). The rail loop from the Main Western Line to the Olympic Park was not part of any metropolitan strategy (Interview Searle 2008). This extension of Sydney's rail system to the Olympic site was rather a necessity, in fact "vital as part of the city's commitment to holding the first public transport-only Olympic Games in modern times" (Palese et al. 2000: 62). The construction of the rail loop and the expected passengers to be transported during the Olympics made it necessary to build a high-capacity rail station at Homebush Bay and revive an old one at Lidcombe. Criticism of the rail loop and its junction as an expected legacy existed. The report by Jacana Consulting found it unlikely that the rail loop would sustain an economically viable and frequent weekday service after the Olympics, because the station would be at least two kilometers from the residential village (Jacana Consulting Pty Ltd 1996). The residential village referred to the permanent Olympic Village, which would have some 6,000 residents. Not only was the train station 2km away, its access would involve a mode change (shuttle bus), and another train change at Lidcombe.

A further transport legacy was to become the new ferry wharf in close proximity to Homebush Bay. It was meant to provide exclusive waterway access to Homebush's stadia for premium VIPs (Department of Transport 1995: 54). The ferry terminal was located at the northern end of the Bennelong road to be operated post-Games as a permanent commuter wharf and incorporated into the Sydney Parramatta Rivercat service (OCA 1995a). It was clear from a transport point of view that the wharf was to be built solely for the purpose of athlete and VIP travel and would not have been built if Sydney had not won the right to stage the Games. As McIntyre (Interview 2008) judges the Olympic bids and their later influence on transportation planning:

> I think bids by their definition need to be aspirational. Things get included that are not particularly useful from a transport point of view. You know, using ferry transport for athletes and officials, I mean, that is a wonderful concept and I am sure it is great at the time when bids are considered, but from a transport point of view it is a complete annoyance, really.
>
> *(Interview McIntyre 2008)*

Originally the wharf was to be built into an existing Brickpit, allowing direct access to the venues at Homebush. However, these plans had to be changed when an endangered species of frogs, the Green and Gold Bell frog, was found on site (Interview Allachin 2008). OCA had to relocate the wharf on short notice. It is evident from this example that in planning for the Olympics uncertainty always hovers over the realization of the plans and there is a clear need for flexibility and contingency planning.

Sydney's road system also had to bear an increased traffic load, and hence key access roads for the bay area and the east–west Olympic corridor were constructed

(Department of Transport 1995). Furthermore, new roads were essential for access to Homebush (OCA 1995c). The site also had to be equipped with 10,000 parking spaces to facilitate the movement of the Olympic buses. As Black (Interview 2008), a professor at the University of NSW and long-standing advisor to the government before, during and after the Olympics Games, recalls: especially the commercial pressure for even more parking was immense at the time. In contrast, opponents to Homebush's road system called for a massive reduction in the 10,000 car park and the road infrastructure of the Olympic site. In their perception, the planned roads with a 60m-wide Olympic boulevard and an urban core that was bounded by four main avenues, each four traffic lanes wide, would induce and encourage car travel to the site in the future (Jacana Consulting Pty Ltd 1996: 9). Especially, the large number of car parking spaces would make travel by private motor vehicle attractive for most events during the year (Jacana Consulting Pty Ltd 1996: 3). Their forecasts predicted vehicle use for 97 percent throughout the year, excluding the Royal Easter Show.

Unfortunately, OCA had no choice in that matter. Stronger emphasis on public transport, biking and walking had definitely been thought of; however, it could not be implemented to the extent suggested. As Black (Interview 2008), also a member of the Olympic transport planning panel assigned by the NSW government, describes the influence of the IOC on planning Homebush's transport: "The Olympic Village, which is at the other side of the creek, is actually, in my opinion, walking distance. So I pointed out that the athletes could walk, until I was told that under the IOC competition rules, they must be taken everywhere on an official bus." This strict but understandable requirement forced the authorities to implement an extensive road network throughout the Olympic Park and provide parking spaces close to each of the 14 venues. The Olympic Games also brought some all-agreed-upon changes to transport within Homebush. Cycle paths and pedestrian walkways were built in the Olympic Park for easier access to the venues. The goal was to increase public transport use and cycling in Sydney as a legacy of the Games (James and Myer 1997). Cycling and walking is fantastic at Homebush, as Grant (Interview 2008) describes in an interview. According to her, the long-term vision for Homebush, to be primarily accessible by public transport, has been fulfilled.

Darling Harbour as the second Olympic center

Early on, Darling Harbour was selected as the second center of Olympic activity. Including the international broadcasting center, the media and some villages for officials, it was located close to Sydney's CBD with the responsibility to host 11 out of the 39 sports (De Frantz et al. 1998; Department of Planning 1995: 37). To connect this center with downtown, further investments in transport infrastructure were necessary. Besides street improvements in the Central Business District, mostly to provide a showcase image for visitors and ensure walkability to the venues, a new monorail opened on 31 August 1997 between Ultimo and

Pyrmont (MLR 2008). Over a stretch of about 800m, it was to serve the re-developed inner-city areas of Darling Harbour linking downtown and the retail core (Department of Transport 1995: 54). From the beginning, the proposal to build a circular monorail aroused major antagonism in the city (Punter 2005). Reflecting on the decision to build the railway, Allachin (Interview 2008) a private developer for Sydney, believed that it was not a good idea. Rather than taking the monorail, residents would walk, and the rail has become purely a tourist attraction. Given important IOC delegates, referees and judges resided in Ultimo, a new modern rail to the retail core might have been an attractive feature to Sydney's bid.

Connecting the airport with the city center

At the time of Sydney's candidacy, the city's main airport was poorly connected to its center, and the NSW government decided to build a rail link and a new highway to connect them. Both infrastructural projects were "seen as a necessity for the successful running of the Games" (Ben et al. 2000: 48). Both developments had been on the political drawing board since the mid-1980s (Interview Black 2008; Interview Searle 2008), yet their entry onto the government agenda was a direct result of the Sydney Olympic bid. In May 2000, Sydney launched its first airport rail link in the run-up to the Olympics for (AUD)$700 million (Palese et al. 2000). The route runs underground, taking passengers from all terminals via Mascot and Alexandria to Sydney's Central Station. Black (Interview 2008) stated that the line was never intended to be a direct airport-to-city-center link; it stopped at various suburban stations. As Daniels (Interview 2008) described the catalytic effect and how Sydney's planners used it to their advantage: "With the Olympics in mind, we realized we needed that link to get people from the airport. But it is really designed to integrate people from the eastern part of the city; it just runs through the airport." The second connection between Sydney and the airport was to become the Eastern Distributor, a major freeway link and tunnel from the southern portion of the city to its center. The highway opened in December 1999 amid vigorous public protests. The political tension surrounding the Eastern Distributor accumulated over the years before the Games. In 1993, the NSW government had promised that no major road infrastructure development would be required for the Olympics (Sydney 2000 1993: 84). Yet in November 1997, the NSW government allowed the building of the Eastern Distributor, claiming it was necessary for the Games (Palese et al. 2000: 63). For critics, it was unacceptable that new road projects were promoted on the basis that they would help to ease Olympic transport congestion (James and Myer 1997: 29).

An explanation for the strong push for the Eastern Distributor might have been that ORTA was committed to providing transport services to the Olympic Family via Sydney's State Road network (Traffic and Transport Directorate 2001: 8). Searle (Interview 2008) recalled that "the Olympic Road Authority always had been very keen on completing the Eastern Distributor." Given that the road was proposed by a private developer at no cost to the government, the project was easily approved and completed before the Games (Interview Black 2008). In retrospect the rail link as well as the Eastern Distributor aligned well with Sydney's strategic transport corridor #1 (Map 5.1). Motivated by the poor traffic conditions during the Atlanta Games, a new Transport Management Centre (TMC) was built for Sydney during the preparation time (De Frantz et al. 1997: 2). The center was intended to keep its function for large events at the Sydney Olympic Park and was to be operated by Sydney's Road and Traffic Authority (RTA) post Games (De Frantz et al. 2000; SOCOG 2001). To feed the center with real-time information, the RTA installed over 350 cameras on the road network by the beginning of the 2000 Olympics (Traffic and Transport Directorate 2001).

Transport performance during the 2000 Olympics

The Sydney Olympic Games took place from 15 September to 1 October 2000. During the Olympic Games, residents, spectators, volunteers and tourists made 38 million trips on Sydney's public transport system. Most of the travel demand occurred on Sydney's extensive rail network. The second largest demand was experienced by the two temporary Olympic bus networks: one serving athletes, the other one serving spectators. Of the 28 exclusive bus routes, 10 Olympic primary routes were solely dedicated to athletes, their equipment and other Olympic Family members. The other 18 were spectator routes, of which nine were to move ticket holders to the Olympic Park and the other Olympic venues. The bus fleet comprised 3,350 buses, 1,000 solely for the Olympic Family, 1,700 for spectators and 650 for sponsors. Officially, public transport captured 75–85 percent of the total traffic demand of spectators, volunteers and the Games' workforce (Bovy 2000; Hensher and Brewer 2002). Parking restrictions and road closures, as well as park-and-ride facilities close to rail stations, supported the free movement of buses. Transport operations were supported by various travel demand management measures, specifically for businesses and other institutions (Traffic and Transport Directorate 2001). A key element to shifting car drivers to public transit was the integrated ticketing systems; each Olympic entrance ticket provided free public transport access to the venue.

During the Olympics, Sydney's temporary transport agencies and transit providers managed to create free-flow traffic conditions; a transport attribute highly appreciated by the IOC. The key to the successful staging of the Olympics according to the IOC evaluation commission were the extensive test events using the Royal Easter Shows as a precursor to the Olympics (De Frantz, McLatchey, et al. 1999; ORTA 1999a: 29). Because of these tests, better predictions of Olympic traffic could be made (ORTA 1998a). In the IOC's assessment, ORTA was providing "optimal transport services for the Sydney 2000 Games" (De Frantz, Palmer, et al. 1999: 53). Post-Games, the IOC was thrilled by Sydney's transport performance: "the Sydney Olympic Games have been a showcase of environmental protection and a tremendous transport success with 100 percent of spectators moved by public transport" (Bovy 2004: 50).

"Under-promise, over-perform" was the motto Sydney had adopted for transport. As Daniels (Interview 2008), a representative from the Department of Planning, describes the media message: "Our focus was on traffic demand management. Before the Olympics we had a program to scare people, so they would take public transport." Olympic planners in Sydney were aware how crucial residents were in riding public transport. Currie (Interview 2008) explains that Sydney was very careful in managing the expectations for the event. Due to the low expectations people had of the transport services before the Games, they were pleasantly surprised by the efficient outcome. Sydney likely adopted this approach due to Atlanta's transport performance. Because Atlanta's friendly request to the public to take transit did not work, Sydney's planners used the "big transport scare."

Transport outcomes and Sydney's legacy

The Games could have been a watershed moment for introducing some key measures that were acceptable to the public and would significantly improve Sydney's transport. Using the Games' momentum for resolving problematic issues, such as traffic congestion, parking policies and integrated ticketing would have been only a few of them. Unfortunately, Sydney rarely took advantage of the momentum the Games provided for transport operations. "We did a lot of good things during the Olympic Games. We should have learned a lot. But the day after the Olympics, everything went back to normal. And we went back to the mess of transport we have in Sydney and it got worse since 2000" states Dobinson (Interview 2008). For this outcome, he blames the state government and their exclusive focus on making the Games an outstanding short-term success. In two instances, however, the Games had created valuable legacies for the city to pursue in terms of transport: experience to stage similarly large events and a cross-town bus system. First, contrary to ORTA's hope (ORTA 1999a), Sydney's Olympics did not change commuting habits in a general sense (Interview Dobinson 2008). However, they did change the culture for travel to very large events (Interview Black 2008). "The Olympics were a deliberate attempt to refocus, away from the car. To the Royal Easter Show at Moore Park, visitors used to arrive 60–70 percent by car. When the Easter show moved [to Homebush], [people arrived] 60–70 percent by public transport. We turned that around" (Interview Grant 2008). Furthermore, operational traffic practices and organizational structures that were used during the Olympics are now implemented for other major events held in the Olympic Park, such as the bus routes for the 2003 Rugby World Cup or the World Youth Day (Interview Grant 2008; Interview Searle 2008). Second, because the concept of cross-town buses worked extremely well during the Olympics, Sydney's planners (Interview Grant 2008, Interview

Daniels 2008, Interview Brockhoff 2008) believe the idea for strategic bus corridors originated during the Olympics. "Some of those bus corridors mirrored very closely those bus corridors that had been used during the Olympics."

In preparation for the Sydney Olympic Games, planners missed important opportunities to secure a strong urban and transport legacy upfront. Only after the Games, significant resources were put in place to continue to leverage the opportunities the Games had brought to Sydney. This short-sighted view is particularly evident in the case of Homebush Bay. Because further developments around the park are taking place, the site poses transport access challenges to the city center. This outcome could have been anticipated and should have been avoided through strategic legacy planning.

Even though Homebush was created to regenerate a long-abandoned brownfield while pleasing Olympic demands, its legacy was still in its infancy shortly after the Games ended. In fact, a vision for Homebush was only thought of after the Olympics had been staged. As an attempt to bring life and frequent use to the site, the government established the Sydney Olympic Park Authority (SOPA) in 2001, which is in charge of organizing events, maintaining the Olympic park site and developing its adjacent property (Interview Daniels 2008). Allachin (Interview 2008), a planner working on the post-Olympics Master plan for Homebush, states that the development of Homebush became a big issue after the Games had taken place. Daniels (Interview 2008) confirmed that there was no real plan on how to proceed after the Olympics. The first developments to occur close to the Olympic Park were residential dwellings shortly followed by commercial office development (Interview Searle 2008). After the Olympics had been staged, the public authorities had realized that with costly public space at hand and excellent facilities, it was best to lease the land in order to develop it. Searle (Interview 2008) argues further: "the strategic plans really adjusted to the fact that the Olympic Games were held in that location." After the Games ended, the NSW government took significant steps to create a viable legacy for Homebush and fulfill the vision OCA had set out to accomplish in 1995: the economic development of Homebush and its surroundings (Cashman 2011).

With no decent transport infrastructure in place designed exclusively for mega event usage, locals feel the burden of this legacy. The Newington area, where the Olympic Village was located, was supposed to be a green model for urban design. Transport-wise, however, it did not live up to its bidding. A current (2008) resident complained about his journey to work: "But what they [government] did was this silly loop-back thing" which required a "20-minute walk and an interchange of trains to reach the city center" – way too long for a regular commute. The movie shown regularly (observed on 15 February 2009) in the tourist center in the Olympic Park claims that currently the Olympic Park receives 8 million visitors a year, of which over 80 percent arrive by public transport. Contrary to this claim seems to be the following statement by a Lidcombe resident, whom I encountered at the Lidcombe train station waiting for the Olympic Park train: "You are the only person I have ever seen on this platform. I go to the Olympic

Park quite frequently." He believed the loop was experiencing an incredible lack of use, but stated it was now being used as a testing ground for new trainees on the train system.

This situation is unlikely to change, because demand for transport services is not high enough. Even though there was a 7.9m wide corridor maintained in the median of the Olympic Boulevard to allow for a future light rail transit system, which could provide a link to Camellia/Parramatta and/or North Strathfield/Darling Harbour, "the development of a light rail transit system is not envisioned in the short to medium term" (OCA 1995a: 21). Almost a decade after the Games, the envisioned light rail has not been extended to Newington, where it could potentially serve new housing development (Interview Black 2008).

The development that has occurred since 2000 in Homebush raises questions about the implementation of the rail loop. With Homebush's large public spaces and beautiful surroundings, planners have argued that the rapid rise of land value driven by commercial developments could have been anticipated. Therefore, appropriate travel options should have been implemented at the time of the bid. Other alternatives for building the railway loop, such as opting to integrate it into Sydney's northern rail system and provide Homebush with a regular CBD service, would have further spurred the development of that area. Dobinson (Interview 2008) described this alternative as more costly, but with the result of having "a loop for the future. [The Games] would have been a great opportunity to do it." Leveraging this opportunity, however, was never the task for which the Olympic transport specialists were hired.

> It was very clear that the rail loop was there to serve large crowds associated with major events for Homebush . . . and possibly for major football events, soccer, etc. that attract 50,000 people. [For that,] special rail services are run on that loop. The answer is very clear: it was to serve these major events and the Royal Easter Show on an ongoing basis.
>
> *(Interview Black 2008)*

A decade after the Olympics, this transport development task is seen in a different light. Allachin (Interview 2008) argued that a better connection, running the rail north to south, routing it through the Olympic Park, and terminating it in Parramatta would have been a much better choice. In fact, his suggestion matches the report on strategic transport opportunities by the Department of Transport (Map 5.1). Daniels (Interview 2008) admitted that the transport link had always been problematic. Due to the loop's poor connection to the main rail network, patronage had significantly declined in ridership and frequency. In the NSW Department of Planning recent Metropolitan strategy, Homebush was identified as one of Sydney's strategic and "specialized centers." Hence, Daniels (Interview 2008) concludes: "We'll probably need to improve transport links to really function and match [Homebush's] potential. We spent a fair bit of money in the first place and now we have to spend a fair bit of money more to make sure that it works."

Conclusions

Sydney's attempt to align its urban vision with the Olympic requirements and create sustainable legacies fell short on the necessary broader investments and decisions, especially in public transport infrastructure. Realizing the vision that Homebush was to mirror Moore Park would have required public transport connections to downtown Parramatta and fostering Homebush's future growth through a more regionally connected transport system.

Unfortunately, the Department of Transport viewed the Olympics as a case study for transport planners rather than an opportunity. "As such, Olympic transport plans may prove to be a valuable case study of the task, which every day faces planners concerned with getting the most out of Sydney's transport system for the least expenditure and environmental impact" (Department of Transport 1995: 54). Viewing the Olympics from such a perspective, the catalytic effect and legacies that such a mega event can bring is neglected. Sydney did use the Olympics as a catalyst in selected examples: to clean up a long-abandoned area by turning a former brownfield into a beautiful leisure park, and enhancing their airport to city-center connection while improving access for local communities to Sydney's center.

However, the long-term urban and transport legacies governmental agencies had promised the public have significant flaws. No thorough transport plans for Homebush's use (besides for events) have been made, and the city is still in the process of developing sustainable solutions, forced to work with the infrastructure that had been implemented during and exclusively for the Olympics. The rail loop allows only poor access to the city center for potential adjacent communities and employers, whereas the extensive parking spaces around the Olympic park allow for car access – a good (yet more polluting) and frequently used alternative. In downtown, the Games brought the impetus for a light rail train, probably to provide access for referees and judges to the main shopping district. This train – barely 800m in track length – is now primarily used by tourists; most locals opt to walk the short distance instead. In retrospect, an integrated planning process building the case concurrently for the Games and the post-Games scenario would have been essential. For that, a separate entity would have been needed, because ORTA was so busy with the delivery of the Olympics, that there was neither staff time nor mental capacity to worry about a sustainable legacy (Interview McIntyre 2008).

As suggested in the introduction to this book, the IOC played a vital role in influencing the creation of Sydney's legacies. Sydney, following Atlanta, was in a special position to prove a car-oriented city could stage transit efficient Games. For the first time, an Olympic transport advisor, Professor Bovy, was exclusively assigned to evaluate transport and advise on potential risks while offering solutions. According to Bovy, Olympic transport functioned outstandingly during the Sydney Olympics. As a lesson for future hosting cities, he identified 12 factors that contributed to the success of Sydney's Olympic transport operations (Bovy

2000). These factors were significant for future bid evaluations and fell under the rubrics of transport policy, transport management and Games' time operations. This early report completely ignored the urban and transport legacies of the built infrastructure and its potential future use. In Grant's opinion (Interview 2008), the IOC had clear priorities in inquiring about transport – none of which included legacy:

> The IOC had an interest in athlete [and] VIP transport as opposed to the visitor and spectator transport. They had a role and interest in setting the standards of what we [ORTA] had to provide for those athletes, and there was this whole hierarchy of who got what level of service. There was this whole contract the government had to provide to the IOC. So [the IOC's] role was really setting the standard of what they were expecting. They did not set quantities [for cars and buses], they set times and levels of service. It was up to us how to fulfill those. We worked on how to best meet that level of service.
>
> *(Interview Grant 2008)*

It is abundantly clear that IOC transport working groups and their advisors primarily sought to ensure that transport during the Sydney Games worked perfectly; the later use of the facilities was left to the city until legacy considerations entered the bidding questionnaire for 2012 applicants. Frequently, this legacy responsibility was and is too much to handle for cities under a tight deadline and with only seven years of preparation (as we will see in Athens and Rio as well). Even if cities understand the need to plan for the city's long-term development, the hurdle of winning the bid while ensuring the city's future development seems too high. Black (Interview 2008), a member of the strategic planning team, recalls the task at hand when a transport and urban strategy had to be identified for Homebush: "Potentially, some commercial development up close. Not housing – that wasn't part of any discussion, certainly *not* the understanding of what we were actually trying to design for." With the IOC's dominating influence on local planning and its power to decide which city is going to win the bid, candidate cities are extra careful in ensuring the requirements are met, sometimes at the expense of viable transport legacies.

6

PLANNING FOR THE 2004 OLYMPICS AND ATHENS' URBAN LEGACY

Athens, like Barcelona, attempted to use the Olympics as a catalyst for urban change. Especially for Athens' transport system, the Olympics provided tremendous momentum to accelerate many of the city's transport projects. After the Games, Athens was left with an Olympic legacy encompassing a new airport, more than 100km of new and modern roads, 90km of upgraded roads, transportation accessible to all people with special needs, 9.6km of metro line extensions, 23.6km of a new trams, 32km of suburban rail, many new parking lots in various locations, new transport management systems, upgraded train stations and a new, ultra-modern traffic management center (ATHOC 2002: 121). However, Athens' government fell short of using the Olympic catalyst entirely for the benefit of the citizens and instead accommodated Olympic peak demands. Compared to previous host cities, the IOC had the most influence on transport decision-making, which resulted in significant project alterations and Athens' urban structure.

Athens' urban and transport history

In the absence of any land-use regulations, Athens has grown uncontrolled, spreading from its port, Piraeus, to its hinterlands and along its coastal areas. This growth has resulted in a colorful mix of incompatible land uses (e.g., adjacent industrial and residential areas) without social amenities and green spaces. Adequate infrastructure facilities to support a growing metropolis were lacking, "particularly those related to mass transport," which resulted in traffic congestion and air pollution (Ministry of the Environment 2009). Historically, planning efforts have led to Athens' current situation. The first comprehensive urban and regional plan for Athens and its surrounding areas was developed after World War II. More or less successful attempts since then, such as the Master plan of 1965 for the greater Athens area and the framework-plan "Capital 2000,"

provided the first legislative context and set goals for the city's further development (Ministry of the Environment 2009: Law 2052/92; OECD 2004).
Transport goals were primarily to expand road systems for an ever-increasing car
demand. Even though more comprehensive plans including public transit developments had been established, no significant progress was made until Athens was
awarded the right to stage the Olympic Games. An academic, who wishes to
remain anonymous (Interview X 2007) believes: "Atlanta and Sydney did not
need the Games, Greece needed them."

Because Athens was designed around an extensive road network, it had
severely neglected its public transportation network and lagged far behind infrastructural developments compared to other European cities in the mid-1990s
(OECD 2004). The random spread of the road network caused major problems
for the operation of Athens' public transportation system that was mainly
composed of buses (Milakis et al. 2008). Two bus companies, ETHEL (buses)
and ILPAP (trolley buses), covering a 5,000km network with 7,000 bus stops
served 1.3 million passengers daily. Additionally, there were two railway lines
operating, connecting Athens to other Greek and European cities. Inner-city
residents were served by a single metro line (Line 1). Since 1869, it has connected
the port in the south with the city center and Kiffissias in the north, serving
300,000 passengers daily.

Calling for a fundamental approach in 1996 to remedy current traffic conditions, the Greek government established the Attiko-Metro S.A., tasked to expand
the current metro line into a city-wide network by 2020 (Attiko-Metro 1997).
Comprehensively, the metro development study of 1996 set forth the key strategy
for Athens' further development based on the current and expected land-use
patterns, projected employment zones and Athens' future transport needs. The
business planning manager for Attiko-Metro, Alexandros Deloukas (Interview
2007) believes that all infrastructural developments up to today follow this
general transport plan. As we shall see in this chapter, this statement does not
quite hold up under analysis.

The metro development study identified three major problems. First, car ownership in the greater Athens region had seen a rapid increase and led to severe congestion in the inner city. From the first transport study by Smith (1973) to the OAS
study (1983) to the metro development study (1996), car ownership had almost
tripled. In 1996, out of 7 million trips per day, 31 percent were taken by public
transport, and 55 percent by private transport (including 6 percent motorcycle and
10 percent taxis). By 2020, the metropolitan transport study predicted a further
83 percent increase in car ownership, if no further measures to control this growth
were taken. Second, the number of illegally parked vehicles was extremely high
(22,000 vehicles per day). This share comprised 45 percent of all on-street parking,
while 97 percent of inner city parking was free. Third, the public transportation
system performed well, but was facing a severe overload (Attiko-Metro 1996).

To solve these urgent problems, the metropolitan development study issued
policy recommendations on how to tackle the task lying ahead. Concrete plans

were set forth to develop a road and public transport network as well as transport institutions that could potentially alleviate the traffic congestion and handle the Olympic transport demands (Attiko-Metro 1996). The road plan called for the completion of four ring roads, in order to relieve inner-city congestion and to serve new developments in the region. Furthermore, upgrading of the congested arterial roads was necessary. In total, the plan called for 187 new road infrastructure projects to be completed by the year 2020 (Attiko-Metro 1996: 24–33). The metropolitan master plan also entailed a massive expansion of Athen's public transport network (see also Maps 6.3 and 6.4). First, it laid out the plan for two further metro lines (Attiko-Metro 1996: 34); Line 2 was to be built from Thevon via Syntagma to Glyfada, and Line 3 was supposed to run from Piraeus via Aegaleo and Stavros to the new airport. Second, an extensive tram network was to span the central city and extend west to Salamina and east to Alimos (Attiko-Metro 1996: 37). Third, the commuter rail network would gain a new circular line, connecting the new airport to the city center and serving more distant neighborhoods of Athens (Attiko-Metro 1996: 39).

Responsibilities for transport development and operations in Athens were divided between the "Ministry of the Environment, Physical Planning and Public Works," which was responsible for the construction of any type of transport, and the "Ministry of Transportation," which was responsible for operating all urban public transport systems and networks (Athens 2004 1997; ATHOC 2004). Because intensive negotiations and coordination were required among the ministries, the Athens Organizing Committee for the Olympic Games (ATHOC) was put in charge in 1998 to lead development and operations negotiations. From then on, ATHOC, in conjunction with the National Olympic Committee (NOC), was responsible for the preparation of the Games, and required to correspond with and report to the IOC. Early on, the IOC gave direct instructions to ATHOC (Alexandridis 2007).

Within ATHOC, a Transport Division was founded "to ensure effective and efficient delivery of transport services to the Olympic and Paralympic families and to the spectators of the Athens Olympic and Paralympic Games while securing optimal financial results and minimum environmental impacts" (ATHOC 2001). The person placed in charge of this division was P. Protopsaltis, who considered a clear strategy and an integrated transport approach to be essential for the Olympics, and for any mega event, to be successful (Interview 2007). Athens Urban Transport Organization (OASA), founded in 1993, supervised the operation of six transport providers and organized the provision of public transport services (ATHOC 2004; OASA 2009).

Athens' candidature and Olympic preparations

Despite the extensive planning efforts of the Greek authorities to prepare their transport networks for the Games, the Pan-Hellenic Federation of University Graduate Engineer Civil Servant Unions (PHF 2001) and independent writers

TABLE 6.1 Athens' three Olympic areas

	(1) OAKA	*(2) Faliro*	*(3) Hellinikon*
Areas in Map 6.1	north	southwest	southeast
# of venues	10	10	7
# of sports	6	4	6

Sources: Compiled by author. Papadimitriou 2004; ATHOC.

argued that the 2004 bid plan lacked feasible technical solutions to obvious problems. Overall, the "planning process [for the 2004 Olympics] seems to have been
addressed very superficially" (Telloglou 2004: 33). For the 2004 bid, ATHOC
decided, partly driven by the locations of existing sports venues, upon two major
competition sites: the OAKA complex, located in the north of Athens, and the
Faliro complex, located in the south, close to the Piraeus port (Papadimitriou
2004). The Hellinikon Olympic complex on the grounds of Athens' old airport
was added as the third Olympic area only after Athens had won the right to stage
the Games (ATHOC 2005). The new three–Olympic-area solution avoided
potential conflicts with existing town planning legislations and provided highly
nucleated centers, in which Olympic sporting competitions were to be held
(Table 6.1).

However, this decision had a tremendous impact on further developments of
the city in the run-up to the Games. In moving the stadiums to Helliniko, the
construction of a public-transport mode became inevitable, because the "IOC
was not satisfied with access being provided solely by private car and put pressure
[on ATHOC] to create a public transport option" (Telloglou 2004: 180).

The Olympic Village (OV) was built in the north of Athens, in Parnitha, a
beautiful area without an industrial history, but lacking adequate transport
connections for the Games. Hence, a road had to be built connecting the village
to Athens' arterial roads. Babis (Interview 2007), the lead traffic engineer working
during and after the Olympics, explains that ATHOC chose this site because it
was close to the highway, Attiki Odos, major arterial roads leading to Athens'
center, and within a 5km radius of OAKA. Furthermore, it was publicly owned
land, easy and cheap to acquire, and, most importantly, the site was not in a
rundown area (Interview Babis 2007). Other Olympic locations in the south had
been considered, but were quickly dismissed; Athens' southern parts had been
industrially polluted and still faced severe environmental problems. For these sites
to be "Olympically presentable, the entire south would have had to be cleaned
up" (Interview Babis 2007). The "Olympically presentable" requirement also
influenced the selection of the Olympic routes: "we also encountered this problem
when planning the Olympic route network. You do not want to have an Olympic
route passing through a bad neighborhood; you want them to pass through nice
ones" (Interview Babis 2007). Further sporting facilities were located along the
coastal lines to the east and west of Athens and its hinterland (Map 6.1).

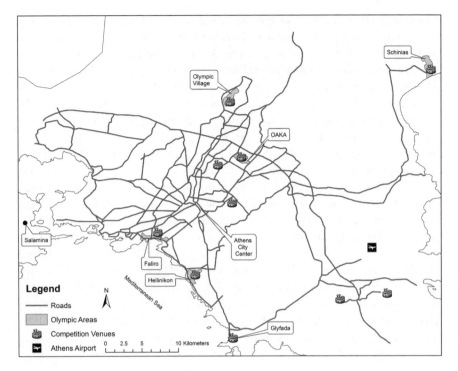

MAP 6.1 Athens' Olympic venue locations

Created by author with GIS (country boundary layer) and georeferencing points according to multiple Google maps. Adapted from Papadimitriou 2004, ATHOC

As Telloglou (2004) and Babis (Interview 2007) have hinted towards, these land-use choices coupled with the IOC mandate set forth the necessity for vast improvements in Athens' road and public infrastructure to fulfill the promises made in the city's bid file. In the IOC interim evaluation report, the IOC members believed Athens to have two major tasks in order to be able to handle the Olympic traffic: alleviating congestion in the inner city and solving the problems with airport access (IOC 1997). Furthermore, the new airport had to be efficiently linked to the city center and the Olympic road network. "In preparing for the Games, Athens had a huge transportation problem at hand, and faced the tremendous challenge of preparing the city for the Games" (Interview Frantzeskakis 2007). Given the existing infrastructure and traffic conditions in Athens, it was clear that the city had to invest extensively in a strong north–south transport connection passing through the inner city, a public-transport mode along the coast for spectators and visitors, improved road connections to all selected venues, and efficient transport management during the Games (Dimitriou et al. 2004). The basic north–south transport connection already existed in the form of the only metro line operating at the time of the bid (Line 1). Conveniently, it connected two of the three main Olympic complexes, OAKA and Faliro. Most other transport projects had to be planned from scratch and the rest significantly altered to meet

Olympic transport needs. Taking on this challenge, ATHOC's planning approach was geared to create a lasting benefit for Athens' future (2004: 6): "the goal was to maximize the legacy of the Games to the Athens transportation system."

Even though the Olympic needs for urban and transport developments were evident, in the first three to four years after winning the bid Athens barely made any progress in fulfilling its candidacy promises (Interview Frantzeskakis 2007). Olympic preparations were stalled by inter- and intra-agency feuds between the ministries, because responsibilities for building and running the infrastructure had not been clearly assigned before the bid (Alexandridis 2007). Other sources (who wished to remain anonymous) claimed in an interview (2007) that construction companies purposely delayed public works in order to increase their payment. As a result, it was feared that the hasty construction would lead not only to increased costs but also to a significant decrease in the quality of the projects delivered (PHF 2001).

These delays did not remain without consequences. Because Athens lagged greatly behind its implementation schedule, the IOC showed Athens the "yellow card." A yellow card means that the IOC feels the successful staging of the Games is in danger. A red card, then ultimately means that the Olympic Games are given to another city (Interview Frantzeskakis 2007). The influence of the IOC on Olympic preparations was seen as a perturbation of the city's concurrent development process:

> Interviewee: "We (Athens) got blackmailed during the Olympics."
> Interviewer: "What do you mean by 'blackmailed'?"
> Interviewee: "We [Athens] were given the yellow card, which means: DO whatever WE [IOC] want!"
> *(Extract from an interview with an academic (X) involved in the planning process, who preferred to remain anonymous, 3 September 2007)*

The imminent threat of losing the Olympics led to an internal restructuring process and replacement of people in charge of Olympic operations. This step was crucial. "If she [Gianna Angelopoulo-Daskalaki] had not taken office, the Games would have gone downhill. It is a matter of the right people at the right time with strong political support" (Interview X 2007). The delay of certain Olympic-related construction projects resulted in the need to complete them in half the time originally allocated. The following years and months leading up to the Opening Ceremony of the 2004 Olympics became a race against the clock, which came to be famously known as the Syrtaki principle.* "Consequently, many compromises had to [be made] in order for everything to be ready on time" (Alexandridis 2007: 9). In the following years Athens rapidly developed its

---

* The Syrtaki is a famous Greek dance, whose rhythm significantly increases towards the end of the song.

TABLE 6.2 Athens' public transport modifications

Metro	Suburban	Tram	Special bus network
Lengthening of platforms for Metro line 1	New suburban rail★ connecting the center of Athens to the airport	New tram line 1[†]	Olympic Family bus network[‡]
New metro line 2		New tramline 2[§]	Spectator bus network
New metro line 3			

Sources: Compiled by author. Adapted from Athens 2004; Transport Division 2003; ATHOC 2005; Attiko-Metro 1996a.

Notes
★ Bypassing Athens in its north, the suburban line intersected with metro lines 1 and 3.
[†] The new tram lines served the coastal venues from Glyfada via the Hellinikon Complex to the Faliro venues.
[‡] The Olympic network consisted of 22 express routes, which served 33 competition venues running on exclusive bus lines called Olympic priority lines.
[§] Splitting in Amina, the second line connected with the city center.

public transport system, as the altered bid file had proposed (Table 6.2). At the same time, all transport agencies renewed their fleets in preparation for Olympic visitors (Interview Kalapoutis 2007; Interview Patrikalakis 2007).

In addition to the massive investments in public transport, Athens also improved its road infrastructure. New highways were built and key roads upgraded to carry the Olympic traffic (Map 6.2). In particular, access from the inner city to the new airport was insufficient at the time of Athens' bid. A new peripheral highway, Attiki Odos, was thought to bring the solution in addition to a new suburban rail line. Bypassing the city to its north, the road was completed shortly before the Games. Without the road, Chalkias (Interview 2007) believes, "transport for the Olympic Games would have been a nightmare." Plans for the highway were in existence since 1992, and the Games provided the catalyst for "finally" building it (Interview Frantzeskakis 2007). Because Attiki Odos provides a connection between two major interstates in the Attica region, it has remained the backbone for the regional transportation network (Interview Chalkias 2007).

Other major road infrastructure was mainly geared to improve access to the competition sites. Therefore, road capacities of future Olympic transport routes were improved. The decision process involved which roads to upgrade: Babis (Interview 2007) described this as follows: "You have, for example, 500 buses during 8 a.m .and 9 a.m. in the morning and they have to go there. What road is it? And at this road we have one lane, because for normal traffic you wouldn't expect more. So you have one lane plus whatever you expect for the Olympics. So what do you do? You build another lane." Especially the Olympic ring road (Map 6.2) was essential in the coordination of the Opening Ceremony convoy. The roads had to have a capacity of 200 buses, lined up by country, without

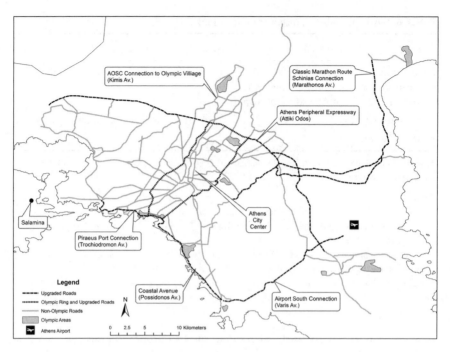

MAP 6.2 Olympic network and road expansion projects in Athens

Compiled by author. Adapted from: ATHOC 2002, 2004, 2005; Papadimitriou 2004)

blocking regular traffic (Interview Chalkias 2007). With all these modifications in place, ATHOC's Transport Division believed that these transport changes would lead to a new culture of transport behavior – Athenians were expected to take more public transport post-Games (ATHOC 2002: 121).

Transport performance during the 2004 Olympics

Coordination efforts among all transport agencies, which functioned virtually as one entity, made the Olympic operation of Athens' transport network a success (Gonzalez et al. 2000). According to Frantzeskakis (Interview 2007), the Games provided for the very first time the necessary impetus for coordination between the Ministry of the Environment, Physical Planning, and Public Works and the Ministry of Transportation. Three further circumstances allowed the realization of Athens' Olympic transport vision: a concrete implementation plan was in place, everyone had a fixed deadline, and the need to fulfill obligations to the IOC provided the necessary pressure. At the same time, flexibility is essential in the run-up to the Olympics: even small changes can have a great impact on the operations and overall performance of the transport system (Interview Babis 2007).

> Transport is very sensitive, very exposed. Olympic buses made more than 10,000 trips. Each one of these trips had a significance of its own. If even

one failed to make it – either it was too late, or had lost its way, or had broken down, or was blocked by a car accident – transport was going to make the headlines.

(Interview Protopsaltis 2007)

Athens had experienced this bitter taste of media criticism in the run-up to the Olympics. The city also had Atlanta as the warning example that bad media reports could cloud the international image of good Games. During Games' time, Athens mastered these challenges with flexibility, preparation and coordination, meeting the guidelines of the Olympic Committee. Like during previous Olympic Games, the Olympic transport vision was to give priority to public transport and to meet the target of being 100 percent on time (Interview Babis 2007). During the Games, a 90 percent public transit share was achieved, while public transport operated under excellent conditions (Interview Frantzeskakis 2007). Additional shuttle lines connected the metro stations to the Olympic venues (Interview Kalapoutis 2007). Free transit for all modes was provided for Olympic ticket holders. In parallel to public transit measures, Athens also had "very good traffic management" (Interview Kalapoutis 2007; Interview Patrikalakis 2007). The planners believe strict enforcement was essential. "During the Olympics, the traffic police [were] almost the Siamese [twin of] Olympic transport" (Interview Protopsaltis 2007). Strict parking controls in the inner city, high fines for using the exclusive Olympic lanes, and crowd management within a 3km radius around the venues were all implemented (Interview Frantzeskakis 2007).

Transport outcomes and Athens' legacy

The positive impacts the Olympics have had on Athens lie definitely in the catalyzing of infrastructural projects that were planned irrespective of the Games. Through the Olympics, Athens was able to significantly improve its public transport system with two metro lines and one suburban rail line (Interview Frantzeskakis 2007). As a result of these investments, the managing director of OASA (in 2007), Konstantinos Matalas (Interview 2007), states that additional riders have been attracted to the public transport system through the Olympics.

The overall assessment of the long-term impacts, however, is disillusioning: "The effect the Games have had on the city is that, up to 2004, there was proper planning, proper implementation, and a very big improvement in everything. After this improvement, unfortunately, nothing has continued" (Interview Frantzeskakis 2007). In 2008, traffic congestion was severe and traffic speeds were quite slow (Milakis et al. 2008). Essentially, Athens had the same traffic problems after the Games as it had before them. Arguably, though, the current situation might have been worse if no public transport infrastructure had been built.

Frantzeskakis (Interview 2007) also admitted that the rushed decision-making on transport projects caused mistakes. By 2007, six or seven studies were underway to completely reorganize the public transport system. Overall, Telloglou (2004: 179) summarizes the common confession of protagonists: "The urban regeneration has been the greatest failure in the Olympic works." Follow-up studies confirm Telloglou's early impression, because Olympic priorities superseded local planning preferences (Chorianopoulos et al. 2010). "Greece did not realize how big of an opportunity it had because of the Games," contends the anonymous academic involved in the Olympic planning process (Interview X 2007). The following sections try to pinpoint these opportunities that Athens seemingly missed in the run-up to the Olympics. The dynamics of the decision-making process are discussed (whenever possible) and the specific influences the IOC had on transport planning are identified.

The biggest chance Athens missed out on was the planning for after-event usage of the facilities it had to build for the Olympics (Interview Frantzeskakis 2007; Kissoudi 2010). Because Athens was left with little time after the early years of political feuds, it had barely four years remaining to prepare for the Games. Given this time pressure, there was no time to think about legacies (Interview Frantzeskakis 2007). Prime examples of this missed opportunity are the empty sporting facilities. Since the Olympics, seemingly no progress has been made on further development of these sites, contrary to the promises laid out in Athens' 2004 bid file. For example, the refurbishment of the old airport, Hellinikon, where the second largest Olympic complex was located, has been delayed indefinitely. In 2007, the old airport was still undeveloped and the plan to create the largest landscaped park in Europe has yet to be implemented. At OAKA, the picture was not much different: dried-up water fountains, empty sites and no tourists.

The loss of integrated urban and transport planning for Athens

After the Olympics Athens was left without a vision on how to proceed and without the coordination structure that had enabled the city to foster its transport projects. Falling back into the existing structures and old habits, the loss of coordination resulted in a halt in projects immediately after the Games. "For Greece, even an event like the Olympics was not enough to change how we do things" (Interview Babis 2007). All Greek planners interviewed expressed a continuing wish that the coordination structures implemented during the Games would have been kept after the closing ceremony. Gaining the right to host the Olympics had provided the city with a common vision through which it integrated the metropolitan planning process and was able to coordinate efforts among all agencies. The power to implement such vision and communicate it effectively to the residents materialized in the run-up to the Games. Protopsaltis (Interview 2007) summarized the change: "our country lacked the background of teamwork and cooperation. The Olympic Games brought that into play, very strongly. And because of it, we were successful." The Greek planners envisioned a metropolitan-wide planning agency in charge of integrating land-use and transportation while coordinating planning efforts across multiple agencies.

The main reason why the coordination efforts cannot become a permanent reality for Athens, so the Greek planners believe, lies in the existing political structures. An interview extract with an academic involved in the Olympic planning process, provides a glimpse into the ongoing political structure of Athens:

> The Ministry of Environment and the Ministry of Public Works are ONE Ministry, it is so obvious that this cannot work. What should be done is to extract the Ministry of the Environment out of the Public Works one and unite the Public Works with the one for Transport. This will never ever happen in Greece. And you know why? Not only because the Environmental heads are morons . . . no, no one wants to lose power in politics, this is why it can not work.
>
> *(Interview X 2007)*

Deloukas (Interview 2007) also supports a ministerial restructuring process; there is little cooperation among both ministries and their planning entities. Communication, per se, only takes place at the highest level between the two ministries (Interview Babis 2007). This lack of coordination originates, so Sermpis (Interview 2007) believes, in the lack of a common policy. The apparent inability to cooperate and integrate is reflected even on lower operating levels. Metro Line 1 was operated by a different agency than metro Lines 2 and 3. Furthermore, planners saw a lack of general resources as an additional hurdle to establish a metropolitan entity (Interview Babis 2007; Interview Kalapoutis 2007; Interview Protopsaltis 2007). A lack of accountability in the existing political system is an additional barrier. All planners agreed that the temporary solution of having ATHOC coordinate between all agencies had been crucial for the operational success of the Games.

Road projects for the Olympics and beyond

Road infrastructure plans deemed to be non-essential for Olympic operations, yet considered necessary for Athens, have been indefinitely postponed. One example is a proposed highway from Piraeus towards Salamina that would reroute traffic currently flowing through the city center. Three-fourths of the road, though in need of repair, already exists and its refurbishment would cut driving time by one hour as well as reduce pollution in the inner city (Interview Frantzeskakis 2007). This project is still in the planning stages, whereas Attiki Odos has been built, even though both were planned around the same time. Hopes for satisfying the ever-increasing car ownership with the newly built infrastructure for the Games are far away. Since Attiki Odos' inauguration, it has seen a constant demand increase (Attiki-Odos 2007).

Since 1985, the Greek government had been emphasizing the need to protect the urban environment along with urging greater use of public transit and the creation of opportunities to walk and bike in the city (MEPPW 1985; Psaraki 1994). A year after the Games, the Athens Olympic Committee claimed that the infrastructural improvements contributed to a significant reduction in atmospheric pollution (ATHOC 2005). However, other voices declared the exact opposite. Organizations such as the World Wildlife Fund (WWF) and independent researchers rated the Olympic performance with regards to the environment as poor, because the environment never had any priority in the planning for the Olympics (WWF-UK 2004; Zagorianakos 2004).

The tram as an IOC-driven legacy

"After the Olympics, let's be honest, the tram has not fulfilled its expectations" said an OASA employee, who prefers to remain anonymous. The cause of this failure was, according to Frantzeskakis (2007), the pressure to finish the tram within two years before the Games. Retrospectively, he admitted that mistakes had been made in the implementation process. Because of those, he argues, the Athenians do not ride it. As one mistake, Kalapoutis (Interview 2007) points out the constant congestion on Athens' roads. With no right-of-way implemented for the tram from the beginning, the travel time was enormous. By 2007, the tram had been given priority. Further problems to this day include the long travel times between various destinations and the city center, and the stops' locations. Those had been planned for access to Olympic venues, for example Hellinikon complex, which were low-density neighborhoods. Now the venues are abandoned, no further development has taken place, and thus the stops became useless for residents. Even though the tram was intended purely as a hop-on and hop-off between the smaller coastal cities, it seems to be an unviable travel option from Glyfada to the city center. Even though riding the tram along the coastal line is a pleasant experience, it takes about 80 minutes from its origin in Glyfada to the city center, while the same distance can be covered in a 30-minute car ride.

The decision-making process for the implementation of the tram reveals the dominating influence of the IOC on planning the tramways. Given that the tram was an absolute necessity for the Games, connecting the venues along the coastal cities, the pressure was on Athens to implement this public transport option, neglecting other options that might have been better for the city (Interview Babis 2007; Interview Frantzeskakis 2007). The anonymous academic involved in the Olympic planning process reports that "the tram line was originally planned to extend to Piraeus, but Bovy and a few others [opposed] that plan" (Interview X 2007). His statement is confirmed by the meeting minutes between ATHOC's Transport Division and the IOC on 12 September and 3 October 2003. The Olympic leaders conducted a "global review of the tram project" and concluded that due to the delays in building the tram, it had become a high-risk project for the Olympic transport system (IOC 2003c). Urgent actions had to be taken, focusing on the *immediate* needs to provide tram transport to the venues. Hence, the planned extension of the tram to Piraeus port was pushed back indefinitely at the time.

Silence is being maintained, however, on a far more important change of Athens' original transport plans. In the metropolitan plan for 2020 (proposed in the metro development study of 1996 introduced earlier), no tram was to be implemented along the large stretch along the coast between Glyfada and Amina. However, due to the Olympics the tram is now running in between Athens' suburbs. The stretch was considered by Athens' planners for the initial tram network, but then filed under non-selected projects, presumably due to its predicted low demands. Once the Olympics came into the picture, the non-selected tram stretch from Glyfada to Amina became essential to connect the coastal venues for spectator transport. Officially, the tram had been intended to serve both the needs of the Olympic Games as well as the future expansion plan of the city. This statement holds for parts of the lines (from Faliro and Syntagma Square to Amina – the selected projects), but not for the extension to Glyfada. The Olympics plainly brought about a change in routes for the trams (Map 6.3).

The original tram development plans (Map 6.3) also had called for a tram extending to the west to Salamina, a city that had been constantly growing but was yet to be integrated in the greater Athens region. Any such tram

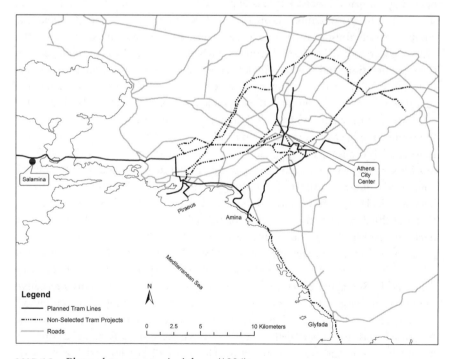

MAP 6.3   Planned tram routes in Athens (1996)

Created by author with GIS (country boundary layer) and georeferencing points according to multiple Google maps. Original Source: Presentation of MDS (p. 60) in 1999, sent by Deloukas via email on 27 February 2009

development has been postponed indefinitely (Interview Deloukas 2007). The decision to develop the tram differently than proposed due to the Olympics, has to be considered in a wider context: given the failure of the current tram performance and the complete halt of further expansions to Salamina, will there be further tram development for Athens and the Attica region? Is this failure a result of the Olympics?

A different legacy and purpose for Athens' metro

Access to Athens' airport is provided by several alternatives: private cars and express buses via the Attiki Odos, as well as the metro Line 3 and the suburban rail, both running on the same tracks. This, according to Babis (2007), was not planned. Even the IOC felt that the service of the suburban rail and metro Line 3 would greatly exceed the demand originating at the airport (IOC 2003a, 2003b).

Despite the IOC's assessment, the double rail service is largely attributable to the Olympic influence on transport planning. Because Athens had significantly fallen behind in implementing a reliable suburban transport option, IOC members and Olympic transport advisors did not believe the suburban rail could be a suitable transport mode for the Games (IOC 2003b). Plagued by construction delays, the operating company Proastiakos was expected to start running its operations only shortly before the Olympic Games. The IOC transport advisors doubted the company would finish construction let alone have any operational experience by the Opening Ceremony. Also OASA wondered whether the suburban rail could perform Olympic functions and in fact complete construction before the Games (Interview Kalapoutis 2007). Hence, Olympic organizers as well as public officials doubted the suburban rail could reliably carry the Olympic peak loads (Interview Babis 2007).

With little time left before the Olympics, a solution was needed – fast, because the connection between the airport and Athens' center was considered "one of the most crucial things for transport during the Olympics" (Interview Babis 2007). A quick fix could provide metro Line 3, if the line could be merged with the mostly completed suburban rail tracks leading to the airport. The meeting minutes between ATHOC's Transport Division and the IOC on 17 September and 3 October 2003 on the global review of the suburban project shed light on the decision-making process: due to the delays in building the suburban rail, the entire project was deemed "inoperable" (IOC 2003b: 2). Hence, the IOC suggested that "during the Games only Attiko metro shall serve the airport with a very fast service from and to Syntagma every twenty minutes" (IOC 2003b: 3). This alternative solution to the airport's access problem was implemented promptly. According to a fax sent on 13 October 2003 by the deputy minister for the Olympic projects, Nassos Alevras, to Gilbert Felli, the IOC's Olympic Games Executive Director, the metro trains and tracks connecting Doukissis Plakentias and Neratziotissa would be able to serve Olympic transport (Alevras 2003: 2). According to Frantzeskakis (2007), the metro agency also was eager to extend the line to the airport to receive a large part of federal funding available due to the Games.

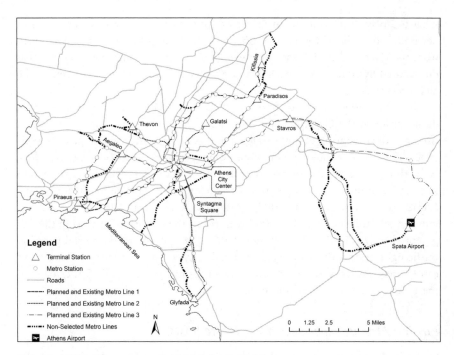

MAP 6.4   Planned metro network in Athens (1996)

Created by author with GIS (country boundary layer) and georeferencing points according to multiple Google maps. Original source: Presentation of MDS (p. 59) in 1999, sent by Deloukas via email on 27 February 2009

The current transport set-up, letting the metro line run in parallel with the suburban rail, was not planned (Map 6.4). Comparing the original plans of the Metro Development Study (1996) with the urban legacy shows significant differences. The preferred option, as calculated by Attiko-Metro S.A., was routing the metro via a northern route converging into the Athens' airport (Attiko-Metro 1996). Hence, the new Olympic metro route – running on a non-selected route further south (Map 6.4) – stripped many neighborhoods located to the north of the current rail lines between Stavros and the airport of a viable public transport option to the airport and Athens' center.

A traffic management center for the future?

The construction of a brand new traffic management center (TMC) has become an essential feature for most cities to manage the large-scale Olympic movements. For Athens, the new TMC was an experiment, just utilitzing the opportunity the Olympics presented to the city. Sermpis (Interview 2007) describes the decision-making process as follows: "With the Olympics, we had an opportunity to do something like this. So, let's build it. But we did not really know what to do with it." The main driver of building the center was, according to Sermpis, the public sector. He describes its attitude as "a kid who can take money from the government, because they were very interested in something like that due to the Olympics." While needed for the Games, the center was to benefit the city afterwards.

Due to the delayed start of construction, the TMC started operating just a month prior to the Olympics. The task manageable with such short preparation time was the operation of 500 variable message signs, which would provide real-time information about ongoing congestion or accidents to Olympic travelers. Even though 1,200 new signal controls were installed, the staff's capacity and knowledge only allowed operating the 500 most important ones during the Games (Interview Sermpis 2007). After the Olympics, the TMC staff was reduced to a minimum and in 2007 was still in the process of exploring the full potential of the TMC. "You witness failures in operations like illegal parking, traffic lights not working in a synchronized way, because the traffic management center does not work properly" (Interview Protopsaltis 2007). By 2012, many of these early issues have been addressed.

Passing Athens' experience on to future hosts

Seminars to prepare cities for the Olympic transport task usually point out the positive aspects, barely warning of the potential pitfalls associated with Olympic legacies. If past hosts do not pass on the true lessons learned and point out potential legacy drawbacks, the same mistakes are likely to be repeated by future hosts. Cartalis (2003), who had witnessed the changes in Athens first hand, made claims in the seminar for the 2012 Applicant City that do not hold up against the evidence presented in this chapter. For example, he stated that the infrastructural investments made during the Olympics "complied with the Master Plan for the development of Athens" (Cartalis 2003: 22). As described in this Athens narrative, the Olympic transport installations significantly differed from the original Master plan developed by Attiko-Metro; in particular, the tram and metro Line 3. Other claims, such as that the Games were aimed to "recover" important sites in Athens, for example, the Hellinikon airport, were apparently true at the time of the bid, but the legacies are quite different; the old Hellinikon airport site has yet to be developed.

Further claims about Athens legacy also differ from the opinions of local transport planners. Cartalis (2003) stated that Athens' planned legacy became a reality with a well integrated multi-modal transportation system with major post-Olympic usage: buses, metro extension, subways and suburban rail. The opposite view is taken by the former head of transport planning for the Olympic Games, Protopsaltis (Interview 2007): "The public transport centre does not really have an integration for different modes of transport, buses, trolleys, trams, or metros. Metro lines 1 [compared to] 2 and 3 are in different organizations, in different entities. This is ridiculous. This is a joke, but it still carries on."

Despite ATHOC's prediction that the Olympic investments could change public transport behavior, transport planners assess the current situation differently. "In most cases we experienced people going back to their old habits" (Interview Protopsaltis 2007). Unfortunately, with no further enforcements in place, compliance with parking restrictions and obedience to traffic laws dissipated after the closure of the Games. "We missed this chance; there was no continuity" (Interview Deloukas 2007).

Conclusions

Athens used the Olympic Games as a stimulus to overcome coordination hurdles among agencies and break political barriers that had hindered public transport development for decades. No doubt, Athens used the Olympics to bring forward discontinuous change to their transport system. In retrospect, they allowed the IOC too much influence over their original transport plans. This was partially the city's own fault. In preparing for the Games, Athens slacked off and lost three valuable years in its preparation for the Games and its future. Because the IOC saw the Games and their success endangered, they had to interfere in the current preparation efforts to push forward viable solutions that were still implementable before the Opening Ceremony. Different routes for the tram and the metro were built that potentially not only disadvantaged communities adjacent to the formerly planned routes but also pushed back other transport infrastructure projects that may never be implemented due to the low ridership on the Olympics-initiated infrastructure that now exists.

Hence, the Olympic Games evidently altered existing urban plans. Contrary to what the government had promised the public, the Games did not solve the city's traffic problems. Indeed, Athens in 2007 faced the same transport problems as it did before: suburban ridership was declining after hitting a peak in 2004 (Interview Deloukas 2007), the tram was officially regarded as not meeting expectations, bus use was diminishing and the inner city was as heavily congested as ever.

In those instances in which Athens managed to align their metropolitan plan with Olympic demands, transport legacies that had been on drawing boards for decades were created in a few short years. Building the Attiki Odos had long been planned by the city and its implementation improved access to Athens' center. Also, metro Lines 2 and 3 could be built in record time due to the powerful Olympic catalyst. Since their inauguration, they have seen a constant demand increase.

7

PLANNING FOR OLYMPIC LEGACIES: LONDON 2012 AND RIO 2016

Since Barcelona's Olympic Games, the International Olympic Committee, host cities and academics have made some progress to better understand the concept of legacies. From refining urban policies to social and environmental impact studies, from the Olympic Games Global Impact (OGGI) assessments to the Transfer of Knowledge (TOK) program, new measures have been set forth in an attempt to harvest the best possible outcomes through the Olympic Games, including urban regeneration projects. To achieve sustainable legacies, the International Olympic Committee has started to place more and more emphasis on potential host cities to justify key legacy aspects as part of their bid files. Being included in Theme 1 as the overall concept and vision for the Games, legacies have become a crucial evaluation criterion for winning the bid. The newest candidature procedures and initial questionnaires of 2016 ask two particular questions targeted to not only understand legacy aspirations because of the Games, but also probe urban trajectories irrespective of the Games. Question 1.4 demands potential host cities to justify benefits of bidding for the Games irrespective of the bid's outcome. Presumably, this question is targeted to identify the catalytic effect the Games could bring to the city and the region. Question 1.5 demands not only a listing of key Olympic legacy initiatives but also their linking with the city's/region's long-term planning and objectives. In addition, host cities are encouraged to describe how these initiatives will "be supported, financed, monitored and measured by all relevant stakeholders prior to, during and post-Games" (IOC 2008a: 66_256).

These initiatives seemingly showed results: London principally won its bid because of its strong and long-desired legacy of urban regeneration. Rio de Janeiro – a first-time selection in Olympic history – has created a committee exclusively dedicated to sustainability and legacy. Writing about legacies of future hosts makes any evaluation or assessment rather speculative and contingent upon further decisions by local policy-makers. However, the following two sections trace the paths of London's and Rio de Janeiro's planning for Olympic legacies, pre and post-candidacy. To foreshadow the analysis, while London is likely to fulfill most of its legacy dreams through the Olympics, Rio will very likely fall into the trap of its many predecessors – with urban realities being a far cry from the Olympic promises made to the public.

Planning London's legacies

London has a rich history of hosting the Olympic Games and will be the first city to stage the modern Games three times (1908, 1948 and 2012). For this first in Olympic history, London's bid committee was required to explicitly describe the legacies the Games would bring to the city. Key legacy questions included how the Olympics would contribute to the further development of sports, people's health and urban regeneration in London and the UK.

Concurrent strategies for urban development and transportation are manifested in "The London Plan[s]," a strategic document that gets updated regularly by the Greater London Authority. The overarching goal of the plan is to make London an exemplary sustainable, world city by fostering the city's economic growth, protecting its environment and cultivating social inclusion. The initial London plan was published in 2004, while proactive work on the document had already started in 2000 (Greater London Authority 2004). The plan identified Stratford as one of 28 opportunity areas for new jobs and housing developments. In particular, transport improvements should be targeted to "support the development of East London, growth in Central London, the Opportunity Areas, and Areas for Intensification, and [promote] better access to town centres and Areas for Regeneration" (Greater London Authority 2004: 48). The East London urban rejuvenation scheme would combine these multiple goals and bring benefits to Stratford's neighborhoods. These include the industrial clean-up of a brownfield, and the creation of livable communities with new job opportunities, public spaces and housing options, summarizes Niall McNevin (Interview 2007), the head of town planning within London's Olympic Delivery Authority (ODA). Investments into East London's transport system would result in a national and international transport node at Stratford. McNevin (Interview 2007), also an advisor to the government during the bidding for the Games in 2003, reviews the potential and current transport infrastructure in and around Stratford: "Ironically, in this part of London, the transportation infrastructure was very good from a macro sense. Micro/miso scale was not very good."

In the 2004 London Plan, the 2012 Olympic Games were already identified as a strategic opportunity (Greater London Authority 2004: 242). The Games would provide the necessary impetus for sustainable regeneration of London's East and be the major catalyst that could transform Lea Valley into a lively community in only a few years. Accordingly, the candidacy file stated that "without the Games, change would still happen, but it would be slower, more incremental and less ambitious" (LOCOG 2005: 23). The development of the 2004 London plan proceeded in parallel with London's assessment of staging the 2012 Olympics. For example, already before 2000, the site of Lower Lea Valley was deemed suitable for the Olympic Games (ARUP 2002).

London's candidature and Olympic preparations

The desire of London to bid for the 2012 Olympic Games first emerged in the mid-1990s when the National Olympic Committee (NOC) of the British Olympic Association (BOA) discussed a potential bid. Following a 1997 manifesto by the British labor party to bring the Olympics to London again, a feasibility study of hosting the Games acknowledged the city's chances. In 2000, the results of this study were submitted to the government as a confidential report. Following ARUP's assessment of the Games' projected benefits and costs in 2002 (ARUP 2002), the House of Commons discussed the great potential of an Olympic legacy for London but required commitment of the country for the Olympic Games (House of Commons 2003: 14): "any infrastructure legacies from ongoing grants or subsidy [should be covered by] the public sector." The House of Commons report was published with extensive recommendations for further inquiry a week prior to the date on which the government had to vote whether to back the bid (House of Commons 2003). With no further ado, the following week London's government officially supported the bid and half a year later, Britain's NOC informed the IOC of London's participation in the race to win the right to stage the 2012 Games.

Around the same time of London's Plan development (2002), the IOC discussed amending the Olympic Charter to include legacy aspects for hosting cities. Following the recommendations of a special IOC inquiry panel for Olympic legacies, the submission specifications for bidding and candidacy cities were altered and a new theme emerged as a valuable evaluation criterion. This Theme 1 was designed to specifically probe the bidding city's legacy aspirations and required a solid justification for them. In detail, prospective host cities had to explain how the candidate city's vision "fits into the city's long-term planning and what legacy is planned for the city" (IOC 2004a: 72). Theme 14, as an evaluation criterion for transportation, had also evolved over time and by 2004 the IOC questionnaire included infrastructure, operations and traffic management policies along with the mandatory governmental guarantees. The questionnaire required multiple tables detailing existing, planned and additional transport infrastructure, parking areas, listing of fleet and rolling stock, and distances and journey time between the Olympic venues in 2004 and 2012 (London 2012 Ltd 2004). After all 2012 applicant cities submitted their initial bid, the IOC published its first evaluation of London's among others in March 2004. The report by the IOC candidature acceptance working group to the IOC executive board suggested London's advance to the candidacy stage, but had deep concerns about traffic despite the promised $30 billion in new investments (Bovy et al. 2004). While the working group acknowledged London's superior transport system, they judged that "considerable investments must be made to upgrade the existing system in terms of capacity and safety" (Bovy et al. 2004: 30_103). Like many of the bidding cities, London's current transport system lacked in capacity to provide "reasonable" travel times. Evaluated rather positively were the planned investments into the Lea Valley, which were coherent with the Games' legacy concept.

Throughout the bidding and candidacy stage, London's media and the 2012 Olympic planning committee sent the clear message to its residents and the IOC alike that the new developments like the rejuvenation of the Valley, the Channel tunnel rail link and Stratford station "were going to happen, anyway, the Olympic Games acted as a catalyst and brought [the projects] forward" (Interview McNevin 2007). London's full bid files were submitted to the IOC by November 2004, focusing on three major selling points in regards to urban regeneration, transport and a sustainable Olympic legacy (LOCOG 2005). The regeneration of the Lower Lea Valley is essential for London to remain an exemplary world city – the Games serve as a catalyst. Massive transport system investments will guarantee smooth travel for all parties involved. The legacies remaining in London will be sustainable in all sporting, developmental and spiritual aspects.

Lower Lea Valley and its promised legacies

Regenerating the Lea Valley site as a post-Olympic legacy essentially meant London's Olympic win, believes Jason Prior, the Master plan team leader (Landscape Online 2011). In particular, London's bid files highlight that "without the Games, change would still happen, but it would be slower, more incremental and less ambitious from a sporting, cultural and environmental perspective" (LOCOG 2005: 23). Locating the primary Olympic Zone in the Lower Lea Valley, which was characterized by degrading infrastructure, community deprivation and high unemployment, ensured massive investments flowing into new facilities, amenities and public spaces prior to 2012.

The Lower Lea Valley regeneration project will create London's Olympic park, the largest urban brownfield rehabilitation program undertaken in Europe in the last 150 years. Within the Olympic Park, the Master plan lays out the creation of six communities with their own unique characteristics: the Olympic Quarter, the Old Ford, Hackney Wick East, Stratford Village, Stratford Waterfront and Pudding Mill. Each community is planned to receive housing options, employment centers and educational facilities along with the necessary competition stadia (Garlick 2009). Also located in this 500-acre park is the Olympic Village, which will be converted into further housing post-Games. Overall, housing will make up around 65 percent of the building space, of which 35 percent will be affordable – a planned legacy for social integration. Lastly, the Olympic Park will benefit the local economy from the full-time jobs needed for Olympic construction, and will give the Lea Valley a new economic base to prosper from in the future, potentially for generations. These, then, were the original legacy aspirations of London 2012.

London's transport preparations and promised legacies

Even though London already had one of the most sophisticated transport systems in the world at the time of the bid, the city's bidding committee promised another $30 billion to be spent on London's rails, undergrounds and roads. In particular East London, and therewith the most important Olympic Zone, was already well connected with four tube lines, two national rail lines and the Docklands Light Rail (House of Commons 2003: 20). By the time of the Olympics, the Park is to be served by a total of 10 rail lines carrying 240,000 people every hour (IOC 2005; LOCOG 2005). A further plus point in London's evaluation, according to McNevin (Interview 2007) is that London's bid has "one of the most compact parks of the history of the Olympic Games. This was one of the sales advantages in the bid." In preparation for London's future and the Olympic Games, most of the capacity improvements are needed for London's underground tubes that connect to the Olympic Park. A 45 percent increase in capacity is planned for the Jubilee Line alone, because it serves the main Olympic facilities like the Olympic Village, the Olympic Park and over 30 percent of the competition venues. Given most of the hotels will be located in London's center, the city also made a push for inaugurating the new Stratford International station with its high-speed Channel Tunnel Rail Link (CTRL) to London's King's Cross station, national cities and European destinations. This connection between King's Cross and Stratford is to cut journey times from the Olympic Park to central London to seven minutes. This temporary Olympic train service will be called the Javelin (London 2012 Ltd 2004). All permanent investments, according to London's bid committee, will leave a lasting positive legacy while fulfilling "the long-term needs of London" (Greater London Authority 2007: 14). Further transport investments include multiple projects, among which are: capacity and access enhancements for Stratford's Regional Station; capacity improvement on the Great Eastern and Lea Valley Line; conversion of the Docklands Light Railway (DLR) to a three-train service; a DLR extension under the River Thames; a new DLR train fleet; additional DLR operational capacity between Stratford and Canning Town; further general infrastructure and capacity upgrade of the North London Line; a new Transport Coordination Centre; further upgrades to the road network along Olympic Routes (the A13 and other, for Olympic travel, essential arteries along the Olympic Route Network); and finally new walking and cycling routes (ODA 2009: 202). Other improvements across London will increase train and metro capacity, again acting only as a catalyst for developments that would have happened irrespective of the Games. In an interview, Wilben Short (Interview 2007), the head of transport for London's organizing committee, states: "the Olympics acted as a catalyst . . . and the capacity of the North London line service will be doubled for the Games and stay that way afterwards."

Operationally, early estimates suggested that Olympic travelers were to increase the demand on London's entire transport system by one percent (House of Commons 2003: 20). Overall, London expects around 7.9 million visitors, and half a million additional trips daily. Similar to Sydney's approach, London's goal is to transport 100 percent of the spectators and volunteers via public transit, and the Olympic Family via the primary Olympic bus networks. London will therefore implement a 235km Olympic Route Network linking the Olympic Park, Olympic competition and non-competition venues (IOC 2005). Overall the investments in roads were minimal, because:

> the delta that London actually has to provide for the Games is quite small. There is very little done on new roads, because London is tight and the policy is anyways not to encourage more road traffic. So for the Games, the only issue really is on how to manage the roads to ensure that the Games' Family and the Games' operational traffic can move very quickly, and that is through the setting up of the Olympic route network.
>
> *(Interview Short 2007)*

According to London's Organizing Committee for the Olympic Games (LOCOG 2005), 80 percent of athletes will be housed within 20 minutes of their events. Short (Interview 2007) further adds that LOCOG, ODA and transport authorities took a very strong stance to encourage walking and cycling in the city for the Games.

Sustainable legacy promises

LOCOG and ODA's commitment was to host the most sustainable Games ever – in economic, environmental and social terms. Legacy, of course, plays an essential role in achieving this goal. In its bid, London promised an enduring legacy for sport, community, environment, economy and infrastructure (LOCOG 2005). As it pertains to transport, the sustainability benefits according to ODA (2010) are twofold: the Olympic Games will act as a catalyst for new and enhanced public transport options across London and for increasing transport capacity in East London to serve the growing needs of residing and, due to the Olympics, relocated communities. With careful attention to the environment, according to McNevin (Interview 2007) all transport infrastructure has to pass the environmental impact assessment as per UK standards. A further attributed benefit, highlighted by the evaluation commission (IOC 2005), is the strong post-Games legacy of the Olympic Park governed by the London Olympic Institute.

After the submission of London's bid files, emphasizing the three legacy aspects of urban regeneration, transport and sustainability, the IOC expert team visited the five 2012 candidate cities between February and March 2005. The Commission released its evaluation report on 22 March 2005 and judged "London would be capable of coping with Games-time traffic and that Olympic and Paralympic transport requirements would be met" (IOC 2005: 78). The evaluation committee seemed to be impressed with the strong legacy aspects that the Olympics would bring to the city, its communities and sports in the country. One example is the positive evaluation of the London Olympic Institute, which would carry on the Olympic legacy to stimulate elite and community sports, Olympic culture, the environment, sports science and research. In regards to urban development, London's argument for the necessity of the Lea Valley's regeneration while integrating the catalytic power for the Games convinced the IOC. The regeneration project was judged to create long-term benefits for the residents across different aspects: economy, job opportunities, housing, educational facilities and sports development. This regeneration project – as stated in the evaluation report – was to take place irrespective of the outcome of the bid. Hitherto, the Games would serve as a true catalyst for East London. Furthermore, during the IOC evaluation committee's visit, the ODA team guided the IOC members through ongoing development projects in London. According to McNevin (Interview 2007) the opportunity for the IOC to actually see the process and construction of the sites was extremely powerful in convincing the IOC of "the certainty of delivery." The transport compactness of the Games, with the Olympic Village located in the Olympic Park, along with the Olympic stadium was also judged favorably. Furthermore, London 2012 promised to be the most rail/public transport oriented Games to date (Bovy 2010: 30). However, in the commission's opinion, the new channel line linking Stratford with King's Cross may be overestimated as it pertains to Olympic transport (Bovy et al. 2004). There were also still concerns about the functionality of the Olympic Route Network. In fact, the IOC transport consultant does not expect the system to work properly unless a 25 percent traffic reduction is achieved (Bovy 2010). McNevin (Interview 2007) counters that at "the time of the Olympic Games during the summer holidays, [there is] an acknowledged dip in transport movements – anywhere up to 20 percent."

In Singapore on 6 July 2005, London was finally elected as the host city, defeating Moscow, New York City, Madrid and Paris. Shortly thereafter the London Organizing Committee for the Olympic Games (LOCOG) was created to oversee the staging of the Games. As its counterpart, the Olympic Delivery Authority (ODA) was to be responsible for constructing all venues and infrastructure for the 2012 Games. Both entities form "London 2012."

Prospects of a 2012 legacy

From the very beginning, LOCOG emphasized the importance of its legacy concepts. Driven by the new requirements of the IOC and its evaluation concept for legacies, the superior planning capabilities of London's planning departments along with close collaborations with the mayor, public entities, private funders and frequent communication with the public, resulted in an ambitious urban development scheme that promises to use the Olympic Games as a powerful catalyst for transport investments and regeneration of London's East. These investments in a broader context should be seen as "public investments in a process of social re-engineering of London's east end" and as means to address a "lengthy period of under-investment in the infrastructure and social develop-ment of the [UK's] capital city" (Poynter and MacRury 2009: 197). By 2011, the Olympic Games have been thoroughly integrated into the London Plan's 2011 version following revisions to the 2004 and 2008 plans (Greater London Authority 2011). The major transport theme in the 2011 version (Greater London Authority 2011: 189) – outlining the strategic framework for the next decades – is a perfect extension of the proposed major rail transport schemes and development opportunities in London (Greater London Authority 2004: 107).

However, the question of pertinent legacies can only be answered once the Games have been staged and several years have passed. Also, the IOC's involvement in the planning process frequently just comes to light after the Games, once the host cities have had some time to reflect on the experience and the planning process. That being said, the IOC's involvement in London's redevelopment scheme have seemed to be minimal due to the superior planning efforts, the holistic Olympic legacy integration with London's strategic plans, and the on-time delivery of the promised tasks leading up to the Opening ceremony. From the experiences of having won the bid and the candidacy, Short (Interview 2007) commented on the involvement of the IOC in the transport process:

> The IOC does not directly influence how the city proposes its bid. The way that a city wants to plan its transport has to be put into the bid. And the IOC has to be quite neutral. So they are not going around telling people on how to do it. However, there is precedence and there are, if you want to bid for the Games, you have to understand, strategies that the IOC are favoring and certainly a more sustainable approach to trans-port is favored by the IOC. Prof. Bovy has published several papers in which he emphasizes mega events being very much public transport based – 100 percent public transport, walk or cycle. Even from those types of documents and strategies, you know that the best way for staging the Games is by focusing on public transport.
>
> *(Interview Short 2007)*

Generally speaking, London's legacy concepts have a positive, well-thought-through image. Their execution, however, has drawn much critique, for example of the escalating costs, modifications to the plan as the Opening Ceremony draws closer, and some promises that simply cannot be fulfilled due to both. Few mention missed opportunities in the planning process per se. One of them, however, is London's crossrail projects that would connect north–south and east–west London through a direct public transit link. Early on, ARUP emphasized the catalytic effect a crossrail would have not only for the Games but also for East London. Before the initial bid, the company urged that "immediate decisions must be taken for the project to be ready in time for 2012" (ARUP 2002), because crossrail would greatly improve capacity and accessibility to Stratford. Despite these recommendations, the government already saw time-related conflicts in completing both major projects: the Olympics and crossrail (House of Commons 2003). In addition, the government entity argued that the exact locations of the crossrail project based on Olympic venue locations kept changing, so that neither the planning process in itself nor the completion time could be relied upon. Even though the mayor strongly believed in 2004 he would have the go-ahead for the project within the year, it was not approved until several years later, while the crossrail project in itself has been postponed, with an expected completion date sometime after 2020 (Greater London Authority 2011). Back in 2007, Wilben Short (Interview 2007) had already stated that the crossrail "will not happen in time for the Olympic Games, so I cannot say it was encouraged by the Olympic Games. However, I think it is overdue, London knows it is overdue."

Planning Rio de Janeiro's legacies

Rio de Janeiro has a rich history of urban regeneration efforts through sporting events (Acioly Jr 2001; Gaffney 2010). Following the Agache Plan in the 1920s, and the Doxiadis Plan in the late 1950s, Rio de Janeiro's development was primarily guided around roads accommodating ever increasing car ownership (Brandão 2006). As superhighways spread across the city in the 1980s, so did the gaping divide between the rich and the poor, further amplified by an economic decline. In response to the escalating impoverishment of large population groups, public spaces were privatized and secluded from general use – only accessible to a small proportion of Rio's population. Dilapidation of public services followed and informal settlements arose across the city, while public and private spaces spread in a disorderly manner across Rio's urban areas (Brandão 2006). In the early 1990s, Rio de Janeiro was a deeply divided and sharply segregated city, known not only for its favelas with high crime rates, but also for its world-famous beaches such as the Copacabana and Ipanema.

The announcement of Rio hosting the 1992 Earth Summit marked the turning point of urban planning that had primarily focused on road users. In preparation for the UN Conference, which was to center around the environment, Rio started reorganizing its public spaces to accommodate pedestrians and cyclist instead. Besides temporary interventions to create such spaces, for example, closing of streets, a few more infrastructural developments were inaugurated to redesign Rio (e.g., the Coast Line Project). In the following years, Rio de

Janeiro's municipality aggressively pursued the public-space creation path and developed a new Master plan in the mid-90s. This new plan laid out a fundamentally different development philosophy compared to previous practices. Instead of road construction, the key issue now was the improvement of living conditions through effective urban management and resource mobilization. Further, the plan proposed a complete organizational restructuring of the municipality accompanied by fundamental institutional changes. The planning system was to be decentralized and interdepartmental working groups were to foster collaboration and communication. The most substantial change in future planning processes, also applied to the development of the Master plan itself, was the involvement of civil society organizations in project preparation and implementation, including neighborhood associations, squatter residents' associations, private and public sector agencies, and professionals, gathering hundreds of representatives to discuss solutions to many of Rio's urban problems (Plano Estratégico da Cidade do Rio de Janeiro 1996). The main goal of the plan was to design policies for urban revitalization focusing on declined areas of the city (Acioly Jr 2001). A further milestone in Rio's new urban management approach was the Rio Cidade Programme executed almost in parallel with the Master plan. It primarily focused on urban revitalization with the aim to create and redesign public spaces (Acioly Jr 2001; Brandão 2006). The plan reported on major alterations successfully implemented in 15 urban hotspots. Combined, these projects covered over 100ha of public space creation, included thousands of new trees planted, new public squares and as many pedestrian kilometers paved as roads upgraded. The goal of all projects was to restore the depleted social life of the city in the hope to drastically improve the quality of life for Rio's residents (IplanRio 1996). The implementation of the strategic plan continued into the early twenty-first century. According to Lopes (2003: 2), a consultant to the strategic planning team, of the "162 prioritized projects and actions, nearly 90 per cent had been performed or [were] in process of implementation."

Rio de Janeiro's candidature and Olympic preparations

Rio de Janeiro's winning of the 2016 Olympics started with its bid for the 2004 Games. As an inexperienced bidder, Rio was not even awarded candidacy status in its initial attempt, because the IOC evaluation committee saw multiple shortcomings in the questionnaire (IOC 1997). These included the lack of Games-suitable transportation facilities to host the Games, "improper" stadia and Rio's inexperience in staging large-scale sporting events. Learning from these recommendations, Rio instead bid and won the right to the stage the 2007 Pan American Games and its Paralympic counterpart in mid-2002. While by no means comparable to the scale of an Olympic Games, it was the stepping stone to bid for the 2012 Olympics. In preparation for the Pan American Games, Rio's organizing committee selected three major sites that received significant upgrades: Barra de Tijuca (as the center with athlete housing),

Maracanã and Copacabana. Most investments flowed into the stadia to make them suitable for the competitions.

Highlighting the already committed investments of almost $2 billion for the Pan American Games, the experience to stage a major sporting event, the new Athlete's village with multiple competition sites and upgraded stadiums, Rio shaped its way to the 2012 Olympic bid. While this time better situated with a more clustered-venue approach, Rio's ambition for the 2012 Games again failed to advance to the candidacy stage (IOC 2004b). Rated extremely low on existing transport infrastructure, the aspiring host city lacked sufficient suburban rail, metro and road infrastructure to connect the four Olympic areas. In addition, the proposed travel speeds appeared unrealistic. Ranking Rio as seventh and well below the required minimum threshold, the IOC judged the prospect of Rio having a functioning system in place as unrealistic due to the vast investments necessary for the Games, the overcrowded road network, and the short preparation time of only seven years. Without abandoning its ambitions to become South America's leading sporting country, Brazil instead pursued the 2014 FIFA World Cup with Rio as the top runner for the final match in Maracanã. Declaring its candidacy in late 2006, Columbia eventually withdrew its bid shortly before the host city selection of the 2007 Pan American Games, making Brazil the only viable host for the 2014 World Cup. The FIFA evaluation commission committed itself to working hand in hand with Brazil, as the World Cup would "far surpass any other event staged in the history of Brazil in terms of magnitude and complexity" (FIFA 2007: 9). In the inspection committee's opinion, Rio de Janeiro had an extensive underground train system and a sufficient municipal bus, minibus and van network. Furthermore the inspection team highlighted Rio's plans to extend its underground rail network and introduce a new light vehicle rail system. The transport infrastructure this time was judged to comfortably meet all travel demands. Furthermore, the promise of 14 renovated and four new stadiums was sufficient for a country that had the best soccer history (winning the FIFA World Cup five times) to successfully stage a world mega event. Included in the bid file was the proposal for the Rio–Sao Paulo high-speed service, which promised to reduce journey times to less than two hours between the two cities. However, FIFA judged the high speed rail to be of minor importance to stage the matches.

Upon Brazil's election to host FIFA 2014, Rio de Janeiro promptly bid for the 2016 Olympics. By then the 2004 bid version had significantly improved, featuring billions of dollars in already committed investments in transport infrastructure, new sporting facilities, and vast public spaces for the Pan American Games and the FIFA World Cup. Furthermore, both sporting competitions would allow for a series of test-events that would aid in preparing for passenger movements of Olympic magnitude. Yet, Rio's bid barely advanced to the candidacy stage when Doha – better positioned than Rio at the time of the bid – was eliminated (Gold and Gold 2011). Again, the IOC working group had major reservations about the transportation system. Not only did the topographical situation pose major challenges to transport operations during the Games, but also the airport would not be able to cope with the additional traffic induced by the Games (IOC 2008b). This critique eventually led Rio's authorities to commit to an investment of $14 billion in urban projects. (If Rio's Pan American Games' history is any indication, this monetary promise will at least increase ten-fold before the dawn of the 2016 Olympics.) The commitments in the bid file promise to produce an enormous urban and transport legacy for Rio, comparable to the scale of transformation of Seoul and Beijing. With such guarantees in place, Rio eventually won the XXXI Olympiad on 2 October 2009 and is likely to showcase a new urban structure and transportation systems with remarkable features.

Rio's four centers of Olympic activity

Rio de Janeiro will feature four zones of Olympic activity surrounding the Tijuca National Park. Building upon the initial structure of the 2007 Pan American Games (Barra, Copacabana, Maracanã), a fourth zone, Deodoro, will complete the four-axis Olympic structure (IOC 2009; Rio 2016 2009). Each of the zones will receive significant upgrades to make them suitable for Olympic demands according to the zone's socio-economic characteristics.

Similar to the Pan American Games, most of the Olympic activity is focused in the Barra coastal zone on the south side of Rio de Janeiro. With 14 Olympic venues, the zone will also feature the Olympic Park, the Olympic Village and the media village including the International Broadcast center. Despite the nucleated set-up, travel distances and times between venues were judged to be reasonable by the IOC due to the high concentration of Olympic activity in Barra. After the Pan American Games, extensive adaptation and expansion of permanent facilities was necessary, for example, six more venues were needed to meet Olympic demands. The Master plan for the Olympic Park lays out the transition from "Olympic Games Mode" to "Legacy Mode" with a particular focus on ecological restoration, the creation of public spaces and security.

The Copacabana zone, rapidly growing and located on the southwest side of Rio, will feature four temporary venues along the coastline of the world-famous beach.

Maracanã, the most densely populated among the four zones, is located in the center of Rio, close to its main port. Given the zone is a stone's throw away from the favelas and the port area is highly industrialized and run down, the entire Maracanã area will undergo major developments and revitalization efforts, including a 30,000 square meter leisure area featuring an amphitheater, a multi-use space, bars and restaurants, and parking facilities. The zone will have four venues including the famous Maracanã stadium, built in 1950 for the FIFA World Cup.

Deodoro on the northwest side of Rio will have seven venues. Featuring the highest proportion of young people across the city, but a lack of infrastructure to support its future development, the IOC hopes to bring significant social opportunities beyond 2016.

Rio's new urban transport network and Olympic operations

At the time of the bid, Rio's transport infrastructure comprised approximately 750km of multilane motorways and major arterial roads, 37km of metro lines, 225km of suburban railway lines, and two airports (Rio 2016 2009). Principally, this was much too little to meet Olympic demands. Because the four zones were not well connected by the existing infrastructure, the candidacy committee proposed a new "High Performance Transport Ring" to connect the four Olympic zones. Circling the Tijuca National Park, the public transport modes include bus rapid transit systems, metro lines and a suburban rail. In parallel, the Olympic lane networks will be implemented on the roads alongside the ring (IOC 2009). In order to accomplish the promised high performance of the ring with its transport speeds, several infrastructural investments were necessary amounting to $5.5 billion, nearly half of the non-OCOG capital investment budget. As per IOC request, these transport infrastructure projects were distinguished between (1) existing infrastructure that needs upgrades, (2) planned infrastructure irrespective of the Games and (3) additional infrastructure necessary for the Games (Comparison between IOC 2008a: 214_65; 2009: 90; Rio 2016 2009: sec1: 103). Some pertinent details follow on each of the three classifications.

(1) In regards to the existing infrastructure, Rio's government plans to implement four massive road projects, upgrades to metro lines 1 and 2, and four suburban railway projects. All the projects are meant to connect Deodoro zone with Maracanã and Copacabana. Also included in the existing budget is the "necessary expansion" for Rio de Janeiro-Galeão Antonio Carlos Jobim International Airport (IOC 2009). This expansion is to increase the airport's capacity from 15 million people per year to 25 million by 2014.

(2) In regards to planned new infrastructure regardless of the Games, major road arteries, metro line 1 and two BRT (Bus Rapid Transit) lines will be built. These projects will cut through the Pedra Branca National Park and the Tijuca National Park, in order to connect the Barra zone with Deodoro and Maracanã.

(3) In regards to additional infrastructure necessary for only the Games, one BRT line connecting Ipanema to the Barra Zone is critical to the Games' operations, so are further road projects along that stretch. These investments will again take away valuable space from the Tijuca national park.

In summary, the bid proposes the construction of over 100km of Bus Rapid Transit lanes and the implementation of 300km of Olympic lanes connecting the four Olympic zones with the airport. The IOC deems both projects as essential for the Games (IOC 2009). In addition, Rio promised to completely renovate its suburban railway system and partly construct or upgrade its metro system. A new rolling stock for suburban rail (70 percent), metro (15 percent) and BRT buses (15 percent) will be deployed. As with previous Games, spectators, workforce and volunteers must use public transit. For the Olympic Family and ticket holders, free public transport is guaranteed and the introduction of traffic management measures is predicted to reduce background traffic up to 30 percent (Rio 2016 2009) allowing an average travel time of 25–30 minutes. On the whole, the IOC judged the transport proposal to be feasible, because implementation and operation of the critical BRT system had proven to be a successful Brazilian transit practice.

The legacy aspect of this Olympic setup also earned valuable legacy points from the IOC evaluation commission. The committee believed the implementation of this transit ring would leave "a significant infrastructure and social legacy for Rio, improving the connection of disadvantaged areas of the city with areas offering employment, recreation and leisure opportunities" (IOC 2009: 61). Once the high performance transport ring is completed, the transport commission of Rio 2016 has estimated a ridership increase from 12 percent to 40 percent with 9.7 million daily trips.

Prospects of a 2016 legacy

Rio de Janeiro will stage the Games of the XXXI Olympiad between 5 and 21 August 2016. Till then, the city has laid out an ambitious plan for infrastructural works expecting to leave a strong legacy for the city's future. The IOC's judgment of that plan is nothing but positive: "the concept proposed by Rio de Janeiro includes an excellent legacy plan" (IOC 2009: 46). According to Rio's government officials and Rio 2016, hosting the Games will accelerate the transformation of Rio, compacting decades of natural urban development into only a few years. Legacies include major infrastructural projects that were already planned for Rio's long-term development (IOC 2009). In regards to transport, the Olympic implementations are in sync with Rio's 2025 Transport plan. Therefore, the IOC's transport consultant Professor Bovy concludes in an interview: "Due to the city's continuing fast growth, there is absolutely no risk of producing a Transport White Elephant" (Ostertag 2010: 17). All of this appears good above the surface.

Below the surface, however, a very different picture of long-term planning, strategic integration of projects into the urban fabric, and broken promises mark Rio's urban development since the advent of planning for the 2007 Pan American Games. The legacy of the Games will likely be nowhere near as positive as London's. For example, if the Pan American Games are any indication on how Rio will implement the promised Olympic legacies, sustainable legacy aspirations seem doomed to fail. According to Chris Gaffney (2010: 18), the Pan American Games have left Rio with an ambiguous urban legacy:

> The production of Olympic constellations in Rio did not deliver the promised transportation infrastructure, did not improve the housing situation

for Rio's poor, did not open new sporting venues in order to develop a generation of Olympic athletes, and neglected promises of environmental remediation while contributing to the generalized capacity of mega-events.

Accordingly, Gaffney (Interview Gaffney 2011b) concludes: "Podemos afirmar que o legado é praticamente nulo" (We can say that the legacy is virtually nil). Unfortunately, similar outcomes are in sight for the 2014 World Cup. The construction of stadia and other facilities for the World Cup are far behind schedule (even later than South Africa was), way over-budget, and implementation is governed through neo-liberal tactics (Gaffney 2011a, 2011c). Common themes echo between the Pan American Games and current preparations for the World Cup: a disregard for the environment, revitalization efforts that drive out disadvantaged population groups and the creation of public spaces for the few who can afford them.

In the spirit of 2016, the prospect of aligning a strategic plan for Rio with Olympic legacies is dim. The latest strategic plan was developed in 1995, principally envisioning the city's future without the Games in the picture. The plan did, however, articulate the goal of making Rio de Janeiro a globalized city. Consequently, the bidding for sporting events, such as the Pan American Games of 2007, and the applications for the FIFA World Cup and Olympic Games formed part of the development strategy laid out in Rio's strategic plans (Lopes 2003). By the time Rio started winning the right to stage international mega events, most of the physical projects as laid out in the Master plan had been completed and the desire for rapid globalization became the center of future development decisions (Lopes 2003; Maricato 2002). With the revised strategic plan, the rights for decision-making effectively transferred from democratic decision-making to Rio's city council. The focus of development was on areas of high real estate value at the expense of the poor and disadvantaged (Pires 2010). In 2009, the plan for Rio de Janeiro was updated again with an even more aggressive focus on globalization. This time, the Olympics played a major role in Rio's development progress. In fact, the official announcement for this urban reform came only a few days after the Games were awarded to Rio. However, the official and completely new strategic plan was not released until late in December of 2010, in which projects were conformed to Olympic needs. It laid out rather minor goals for improving living conditions in the favelas, but major ones for tourism and real estate (Pires 2010; Prefeitura do Rio de Janeiro 2010). If all projects were to be implemented as proposed, Rio would have exclusively focused on attracting international capital, hyped real estate developments and become a global city at the expense of the poor. Rio, as a more and more segregating Olympic city, had effectively ignored its original strategic plan premises to support integration and reduction of poverty. Instead Rio's government created a new urban structure excluding the favelas and in particular Rio's deprived northern and western suburbs – those that are in desperate need of investments (Gaffney 2010).

The promise of implementing "sustainable transportation legacies" (Rio 2016 2009) in particular, seems to be way off track. Despite the IOC's assessment, transport white elephants seem likely: one prime candidate is the transport link connecting the military zone in the center of the city to the Olympic Village, on which likely little transit demand will likely remain post-Games. Also, the Olympic transport proposals come at a high cost for the environment, because "the majority of Rio's Olympic installations will be built on the wetlands of Barra de Tijuca" (Gaffney 2010: 24). Olympic construction as planned will not only disturb the natural habitat, destroy precious wetlands and lessen water quality, but also pass through and destroy existing neighborhoods.

When and how many of these projects actually will become reality is in jeopardy. By examining the transportation networks planned for the 2007 Pan American Games, Gaffney (2010: 10) notes that "all of the transportation lines shown in the map were pre-existing. The proposed metro, bus, train, and ferry lines never went beyond the planning stage." Furthermore, as with the FIFA bid, the Rio–Sao Paulo high speed rail made it into the 2016 Olympic bid files. The line, however, is already plagued by delays and unlikely to open neither before the FIFA World Cup in 2014 nor the Olympic Games in 2016 (Green: 19 July 2011). That being said, the previous chapters leave little doubt that essential transport connections will be functional by Rio's Opening Ceremony.

At the end of 2011, transport projects for the Olympics are in continuous planning flux. Most original plans are being shoveled around to accommodate not only IOC needs, but also the short time horizon before the Opening Ceremony. Because projects are faced with extensive delays, alternative solutions that can still be implemented are invented. For example, the metro is undergoing ever changing plans while construction work has already started. The newest plan now is to have the metro run to the closest point in Barra de Tijuca and let passengers transfer to other surface modes – this was not in the original candidacy files. For now, the IOC has approved all changes in transport plans.

Conclusions

London and Rio de Janeiro are on opposing ends of the spectrum in planning Olympic legacies, while moving their urban agendas toward the betterment of the cities' future.

London's planning approach for the Olympics went hand-in-hand with the development of London's strategic plan. While the development of new Olympic requirements and potential Olympic Charter alterations were carefully tracked by London's Olympic bidding entity, parallel sustainable implementation accompanied the development of the London Plan. Overall, the Olympics did not fundamentally alter the vision London had in mind for its future. For a long time, the city's government wanted to redevelop and rejuvenate its eastern parts, with Stratford as the center of investments. Furthermore, there had been a long-standing aspiration to build an international rail connection to London's metropolitan area. Both visions could be achieved through the Olympic Games. However, something's always "gotta give": even though the crossrail had been on the host city's drawing board for years, the city was not able to leverage its development through the Games, because it was not necessary for Olympic movements.

For Rio, strategic planning has been guided since the late-1990s by the narrow demands of mega event needs. The Pan American Games, the FIFA 2014 World Cup and the 2016 Olympic Games will have dictated the development of Rio de Janeiro for almost two decades by the time of the last closing ceremony. From the outset, Rio de Janeiro chose Barcelona as a model, with direct inputs from consultants from Barcelona into the development of Rio's strategic plan (Acioly Jr 2001). Despite Rio's arguably smart attempts to copy the successful regeneration model of Barcelona via the Games, the key ingredient for success – a strategic plan and vision for Rio de Janeiro's physical urban development – was missing. From a superficial point of view, both Olympic hosts show similar mega event structures that might lend themselves for direct copycat versions: four centers of Olympic activity, a high-capacity transport ring, a highly industrialized and underdeveloped port, and a culture passionate about football (soccer). The urban and transport legacy of three internationally renowned mega events will leave Rio de Janeiro with an urban structure that is set up to stage and accommodate the needs of other mega events, but not for an integrated urban fabric that is built upon the principles of urban sustainability and the needs of the populace.

Planning Olympic legacies truly depends on the city's planning goals. Many scholars have found that the "pulling off the shelf" ideas of pre-existing plans is a common phenomenon in prospective host cities. If such plans exist with a vision for the city in 30 years, the Olympics are one of the most powerful catalysts a city could ask for. However, if the plan and the vision do not exist, the Olympics can fundamentally serve as a distortion of the urban structure and legacies are beneficial to the very few people whose interests the Games serve.

8

TRANSPORT DREAMS AND URBAN REALITIES

Past Olympic host cities pursuing the IOC's transport dream of highly efficient, integrated, reliable and all-around excellent transport services during the Games were frequently confronted with unexpected urban realities. Contrary to the hopes of planners, governments and residents alike to use the event as a catalyst for a permanent increase in public transit usage, operational implementations were mainly of a temporary nature and led only in a few instances to long-term commuter benefits. Infrastructural improvements showed mixed outcomes. Some remain as "white elephants" (high maintenance costs, low ridership) in host cities, stemming from the lack of alignment between their own urban vision and the Olympic transport dream.

The exceptions to this outcome originate from those city planners who took a strategic view and leveraged the Olympic Games – sometimes very boldly against the preferences of the IOC. If the host city is not prepared with a very clear vision of its desired urban form regardless of the Games, if it does not cleave to that vision during its bid both in land-use and transportation choices to fulfill the IOC mandate – then the outcome for the city is going to be no more than expensive short-term changes to satisfy the peak demands. In this situation, there would be minimal positive long-term impact on the transportation system and urban structure of the city. In the opening chapter of this book, I posed three key questions that guided me through the exploration of legacies in six Olympic cities. Here, I return to those questions to recap my findings and discuss implications for future hosts – and more importantly, for you as a reader.

Is the urban planning process altered due to Olympic requirements? If so, how?

For past host cities, urban planning goals were often constrained by Olympic requirements. As a result, the interplay between the IOC's requirements and metropolitan planners' visions resulted in a change of the host cities' original urban and transport goals. While host city governance and local interest groups sought to push their own agendas, the big picture is that the Olympic requirements frequently supersede local pressures *unless* planning processes have been well established and integrated into the city's strategic plan well before the city bids for the Games. Academics, researchers and the IOC have argued that mega events like the Olympic Games are a catalyst for urban change. Catalysts, in their true sense, speed up developments, but do not alter them. A comparison between the original city plans published before the Olympics and the actual modifications undertaken in the run-up to the Olympics, however, reveals the Olympics to be more than simply a catalyst. While the Games certainly were a powerful catalyst for some pre-planned infrastructures (as per existing urban master plans pre bid), in other instances, they created new infrastructures that were never part of the city's Master plan irrespective of the Games. The discrepancy between both shows the degree to which the Olympic Games have influenced existing metropolitan plans, often by way of IOC requirements of host cities' new Olympically-driven priorities (Table 8.1).

Barcelona's approach of taking the Olympics as an opportunity to redesign their city was exemplary. Part of this success was that planning began very early – visionary documents reach back to more than 20 years before the Games. In complying with the IOC's requirements for stadia, the city was forced to add two further areas of refurbishment to the original city plans. The key word here is "add"; while the city complied with the IOC rules, it never gave up its other 10 areas of refurbishment nor did it change its overall vision for Barcelona. In terms of urban form, the requirements of the IOC gave an opportunity to the planners to expand Barcelona's center and redesign the entrance gates to the city. In terms of transport, which followed from the land-use choices, the Games acted as a catalyst for creating the ring road (on the drawing board for 30 years) and other road expansions in the inner city. These were primarily pushed forward by the host city's desire to present the IOC delegation and athletes with first-rate traveling conditions. In contrast, investments in public transport improvements for Barcelona's residents fell short because the existing transit infrastructure already served the most important travel routes to and from Olympic stadiums. Hence, other potential metro expansions were delayed till after the Olympic Games.

Atlanta's example shows that even with minimal resources, the Games can be staged. The pre-condition for such an achievement was that plenty of existing stadia were available for use, and the key transportation approach was to overlay

TABLE 8.1 Transport developments during the run-up to the Olympics

	Not in urban Master plan = deviates from	*In urban Master plan = catalyzed through the Games*
Barcelona	Escalators Subway stop: Villa Olympica Funiculars	renfe rail tracks Rondas
Atlanta	Exit I-78	MARTA extensions HOV lanes IVHS
Sydney	Rail loop Monorail Ultimo-Pyrmont Roads around Homebush	Eastern Distributor Rail to the airport
Athens	Tram (different route) Metro Line 3 (different route) Roads around Olympic Ring	Metro Line 2 Attiki Odos Suburban Rail

Source: The author.

Notes

The table distinguishes infrastructure developments that had been made in the run-up to the Olympics (completed between the election of the city and hosting the Olympic Games) according to Olympic inspired changes, and catalytic changes. Olympic inspired changes include those that were significantly altered or built solely because of the Olympics (deviates from, not in urban Master plan). Catalytic changes (as used in this context) are those Olympic infrastructures that appear in the metropolitan plan prior to winning the Olympic bid. The bid turned host-city contract likely denies cities the chance to grow smarter during their preparations (seven years), because they are locked into the promises they made prior to winning the Games.

the Olympic passenger flows on top of the daily commuting of Atlanta's residents. Theoretically, this approach minimizes investments and maximizes the benefits for the residents' daily commute after the Games. Practically, however, this approach can – and did – result in severe traffic congestion during the Olympics. On one hand, Atlanta remained in control when staging its Games and scarcely wasted any money on unnecessary transport infrastructure. On the other hand, its reputation for transport remained tarnished even years after the Games. If the ACOG had paid closer attention to media transport, then the global press would have likely been more positive. Overall, Atlanta planners could have used the Games to further their transport vision: an Atlanta that was known for its modern multi-modal transport system. Without the needed funds or public support, such a vision could not be implemented and Atlanta remained car-dependent.

Sydney dedicated Homebush Bay to sports and post-Games to commercial developments. In the wake of the Games, Homebush experienced a tremendous development boom in its surrounding areas, which could have been anticipated. Because Olympic planners did not pay much attention to the period after the Games (or predict the boom, for that matter), many opportunities were missed and transport legacies were built that future planners for Homebush would have to struggle with. The rushed decision to build the train loop resulted in a poor connection to downtown, which consequently is rarely used on any regular basis. To keep Homebush lively, the Sydney Olympic Park Authority organizes frequent events at which ample parking space allows easy and convenient access to the park by car. Even though citywide, the Games did not have the impact they could have had, Homebush has become a thriving center for new communities and commercial developments.

Athens, on the one hand, used the Olympic Games to push forward numerous public transport projects, many of which – so planners and citizens agree – would not have been completed were it not for the Olympics. On the other hand, while extraordinary efforts were undertaken to finish the promised infrastructure for the Games on time, the IOC's pressure to implement certain routes using specified modes compromised the transport plans that the city had envisioned for its future but ensured the city could actually stage the Games. This change in plans may now result in a complete dismissal of other planned projects due to the success or failure of the differently implemented train, tram and metro routes.

Future hosts like London and Rio de Janeiro will have to adhere to the IOC's legacy requirement, which obliges potential hosts to justify their planning approach. Multiple questions now ask which types of sustainable legacies will result, how Olympic infrastructures will be integrated into the city fabric, and how the legacies will fit into the concurrent long-term strategic plans of the city. Through the required conceptual planning approach, it seems future hosts have better chances in leveraging opportunities, as pre-planning for post-event usage has become an important evaluation criterion for the bid. However, a thorough assessment on whether legacies were intended or resulted due to Olympic planning can only be done years after the Games; once the true legacies have emerged. Concurrent planning efforts for both, London and Rio, provide a first indication on potential legacies and the host cities' urban realities.

London's planning approach seems to be exemplary from an alignment perspective. The London Plan and its updates over the preceding Olympic years integrated early visions for the Olympics. In addition, the approach has located venue clusters according to existing facilities and degraded urban spaces in dire need of urban renewal. London's political advisories for urban planning and Games planning have closely monitored the process by which the IOC developed and eventually implemented the sustainability and legacy requirements in the bidding questionnaires. Accordingly, local advisories integrated these new requirements into the bid and Olympic-related London plans. This careful alignment might lead to a superior legacy for London. A precondition for such legacy success, however, might be that the city already has extensive mega event experience, a long-standing, metropolitan-wide planning agency and a superior transportation system.

Rio de Janeiro's win to stage the 2016 Olympic Games occurred at the cusp of moving from a local to a global city using a series of sporting events as a rapid accelerator towards globalization. In the process, however, planning for the city seems to have become a quest for providing optimal conditions for these mega events rather than a well-thought-through approach of urban renewal, regeneration and rejuvenation. The latest strategic document dates back to the mid-90s, when IPlan Rio offered an insight into a series of targeted urban projects. The comprehensive Master plan, published around the same time, focuses on integration and cooperation, the exact opposite of Rio's concurrent mega event planning processes. The next strategic plan, "2016 and beyond," indicated that thinking and rethinking of plans will start with 2016 – once all mega events have been staged. With this planning approach, it is likely that Rio will end up with an urban structure that was – for over two decades – purely driven by sporting events.

What role does the IOC play in urban development and the creation of transport legacies?

The IOC has a specific set of goals that it wants to accomplish with regard to transport that are the same no matter which city the Games are staged in. Across cities, the IOC's purpose is the same, its time horizon is the same, and its prescribed means of transport are the same. These prescribed means and preferred mobility and accessibility options can have a powerful influence on the urban and transportation plans of the cities, as evidenced by some of the case studies; cities changed their planning priorities, their urban plans and ultimately their long-term development to accommodate Olympic needs. Hitherto, the Committee holds significant power in creating Olympic legacies. It can actively alter pre-Olympic urban plans and does catalyze those infrastructural projects that are essential for Games operations.

This role description, however, is only part of a larger story. Essentially, the IOC is taking a business-like approach to ensure cities stage successful Olympics. Host cities, in return, are responsible for planning Olympic legacies. While the IOC is relatively unfamiliar with the bidding cities' political climate, cultural history and internal planning processes, the Committee can only provide requirements and guidelines on how to make an Olympic Games successful. Furthermore, it can only evaluate what the host cities choose to present in their bids. In contrast, bidding cities themselves are much more attuned to their own urban environments, know how their planning system functions, and intrinsically are aware of potential glitches that may prevent planned legacies from becoming reality. In the past, host cities seem to have overpromised legacies to win the bid and subsequently regretted some of their decisions. The winning legacy stories are ultimately written by host cities that manage to align their urban visions with the Olympic catalyst.

Arguably, host cities experienced at least some pressure in providing appropriate Olympic transport facilities. Over the past decades, the IOC has gained experience with previous hosts and has started creating requirements for host cities to meet. Cities are encouraged to comply with these requirements and commit to adjustments of their transport systems in their bids. Through the introduction of the Transfer of Knowledge (TOK) program, Olympic organizers are ensured a systematic approach so that no knowledge gained through previous host city experience is lost. Due to the TOK program, the IOC and cities underwent a steep learning curve in preparation for the Olympics. Because new lessons emerged, bidding cities now have to comply with new rules.

As experience and best practices accumulated and knowledge management became more prominent, the IOC's requirements grew substantially and became more stringent as Olympic sports proliferated. While the requirements seemed to be rather lax in Barcelona and Atlanta, the following host cities, Sydney, Athens, London and Rio, were faced with more demanding transport adjustments. Particularly since Atlanta's Games, the IOC has adopted a more rigorous approach. The Committee has ensured athletes reach competitions and are granted secure rights of way. Media transport was revamped because the press was crucial for broadcasting a positive image to the world. Therefore, in later Olympic Games, they received dedicated buses and cars running on the primary Olympic network. Furthermore, extensive mass transit systems were needed to cope with the peak passenger demands for the Olympics. In 1999, an IOC transport advisor (Professor P. Bovy) was named to oversee transport preparations. Given the success of the Sydney Games in regards to public transit, the IOC transport consultant set forth that "only very high performance public transport has the necessary capacity to meet the very peaked and high traffic demands of mega-events" (Bovy 2004b: 20). He got even more specific in suggesting that the spectator traffic shall only be supplied by high capacity *rail* transport, because it can deliver traffic demand peaks of 50,000–200,000 persons/hour (Bovy 2004b: 23). Hitherto, all following Games were primarily reliant on rail systems as a means to transport spectators, volunteers and the work force. Only Rio provides an exception to the rule due to the city's extensive Bus Rapid Transit (BRT) system.

An alternative explanation of why Rio, Athens and Sydney experienced more alterations in their pre-Olympic planning, might be that the IOC did not anticipate any problems in London, Atlanta or Barcelona, and hence did not interfere. With that explanation, the interpretation of my findings would be that the degree of change cities are forced to make is directly correlated with the city's timely preparations and how well those are suited to make the Games a success. Basically, the more the IOC anticipates problems with the current strategy, the more it will intervene. Even though staging the Olympics empowers metropolitan governments to expedite the decision-making processes, the IOC's influence on local transport plans has not been well received by local planners: "the IOC is not really interested in [long-term impacts on the city]; all they are really interested in is forcing the Games holder to do a good job" (Interview Currie 2008). Because till 2004 the focus of the IOC had been exclusively on the

Olympic event itself, without taking the city's future development into account, poor legacies have remained in cities. For example, the underused railway loop to Sydney's Olympic Park. It was a temporary solution sought by Sydney to accommodate the peak passenger demands, while planners now struggle to realign it with Homebush's growth (Interview Dobinson 2008).

With the introduction of the sustainability and legacy justifications in bids (IOC 2004), the previous trend of unfulfilled planning goals might turn the fortune of host cities to create valuable urban and transport legacies. Presumably, the IOC has become aware of the consequences their stipulated requirements can cause. Therefore, the committee added legacies as an evaluation criterion and favored those bids that matched the host city's Olympic plan with the host city's vision to create sustainable legacies. London and Rio are the first to experience the new requirements' outcomes. In regards to expected legacies as promised in bids, however, we should be cautious in examining the host city's role in creating them. Evident in previous bidders is the almost immediate agreement to comply with and exceed IOC recommendations (e.g. Athens' and Rio's subsequent bids). If this behavior continues, host city planners might overestimate the legacy potential of their Olympically stimulated urban changes. In short, the focus on winning the bid will dominate the planning process. As Daniels (Interview 2008) summarized the principal conflict: "Before the Games, the focus is on winning the bid, and once you've won the bid, the focus is on delivering the event."

Given the IOC's role, how can host cities plan for valuable Olympic legacies and aspire not only to host the best Games ever, but also retain the best legacies ever?

Future host city planners have to be very careful about using the Olympic requirements for the future growth of the city. This requires long preparation time and extensive planning efforts, in which the land-use choices and transport investments are considered in light of post-Games scenarios. Only then can the Games become a powerful tool in leveraging desirable discontinuous change. If the Games are used strategically, they can catalyze not only infrastructure but also transit and traffic operations that can significantly improve urban transport in the long run. However, many opportunities in the wake of the Olympics have been missed because no actions were taken to ensure the sustaining of beneficial transport developments the Games had introduced to host cities. After the Games were over, cities had lost the vision, the agency wielding temporary power, and the special coordination mechanisms among transport modes. With this loss, measures were revoked that could potentially have alleviated traffic and significantly improved transport conditions in the city – a key promise made and reiterated to the public in the run-up to the bid. To forestall this loss, cities should seize the opportunities the Games present by using the Olympic momentum to coordinate transport agencies, to give priority to public transport and to encourage citizens to use public transit permanently. Hoping for such change, however, is worthless without thorough post-Games plans and guardians who defend them.

Alignment of planning goals: comprehensive planning and a 20-year horizon

A successful outcome of the Olympics depends on good strategic planning that minimizes the additional efforts necessary to stage the mega event. Hence, planning should aim for the alignment of the Olympic requirements with the strategic plans for city development. The key is to identify win–win elements creatively by testing Olympic scenarios against existing metropolitan plans. It is essential to defend and hold aloft the identified synergies and to use the Games as catalysts for desirable change. Cities implemented various changes into their public and private transport systems with the goal to provide access to Olympic venues during Games time and to ensure a beneficial legacy for the city after the event. This alignment usually is reflected in the degree to which cities managed (1) to catalyze urban and transport projects manifested in their strategic plan, (2) to fulfill their urban vision, (3) to meet the goals they had promised the public and (4) to leverage the opportunities the Games brought.

A key problem cities encountered in the past is the short planning horizon for transport: while the IOC advocates about six years (see Bovy 2004a), the widespread opinion among Olympic planners is that three years of transport planning is sufficient. Dimitriou et al. (2004: 11) pointed out that "early planning (3 years in the case of the Athens 2004 Olympics) is necessary for thoroughly examining all potential operational and planning aspects of the transportation system." To stage an operationally functional event, this might well be the case, but to ensure sustainable and more comprehensive planning (such as using the momentum of the Games to inaugurate an integrated ticketing system), much more time is needed. Therefore, holistic, integrated and comprehensive planning for sustainable legacies has to start *years before the bid* files are submitted to the International Olympic Committee and has to be geared towards after-event usage. In fact, the best urban legacies were gained by the cities that planned the longest and were able to take existing Master plans "out of their pockets." Barcelona, for example, started planning for the Games as early as 1977, whereas the bid was due in 1986. In sum: the closer the Olympic Opening Ceremony, the greater the pressure to meet Olympic demands. With such tremendous pressure on planning, creative alternatives guided by a vision are hard to implement.

All cities claimed in their bids to have a post-Olympic vision in mind by leveraging the Olympics for a locally attuned outcome. Yet, past hosts leveraged only physical transport changes – some more and some less successful – and neglected operational ones. Regardless, there was one ingredient to make plans become reality: pre-Games commitment for post-Games implementation. Yet neither political commitments nor a coordinated strategy beyond the event were common features in the pre-Olympic planning before 2012. As an exception to the rule, Sydney established the Olympic Park Authority as an entity to oversee the further development of the Olympic Park and attract visitors to the regenerated area. Unfortunately, because the Park Authority was established only in 2001, it struggles with the uncoordinated legacies the Olympics have left. As two Sydney interviewees believe, the city now needs to invest even more money into transport, if Homebush is to prosper (Interview Allachin 2008; Interview Dobinson 2008). London has taken this lesson to heart and established a permanent Olympically focused authority well in advance of the 2012 Games. Similarly, Rio promised to inaugurate an authority meant to last years past the Games and has already put a Master plan in place post 2016 (Prefeitura do Rio de Janeiro 2010).

During the bidding, candidacy and preparations period an entity has to review the expected legacies periodically. This legacy entity must remain in cities beyond the mega event to ensure their proper realization. It also has to adapt the legacy vision to unexpected changes. As an example, Sydney's ferry system, planned for the Olympics, had to be revisited after an endangered frog species was discovered during construction of the ferry wharfs. Similarly, Athens metro came to an abrupt halt when ancient artifacts were found during the digging work. These two examples demonstrate the need for a flexible strategy, but also the need for additional time buffers to ensure meeting the fixed Olympic opening deadline and retaining beneficial legacies.

Political guardians for intended legacies

Guardians are key for establishing a long-term transport agenda and keeping the good transport practices experienced during the Games. These guardians are strong leaders who leverage the mega event for the city's benefit – sometimes against the IOC's wishes – penetrating different levels of governments and jurisdictions. They are influential figures, who buy into the city's vision, stand up for it against the IOC or public resistance, and who are so situated as to be able to influence plans and projects over a long period of time. These guardians can be leading people in a business community (e.g., CEOs of airports) and public figures who remain prestigious over time (e.g., mayors or former mayors, people holding positions in the chamber of commerce or people who are executive directors in public transport).

Besides this major task, these guardians should secure buy-in from political parties during the bidding stage. Contracts have to be negotiated with government officials and signed on the basis that if temporary measures prove successful during the Olympics, politicians will support their continuation thereafter. Setting aside a budget for these or any additional transport modifications necessary for future maintenance of infrastructure, and for the permanent execution of operational tasks, are essential features in designing these contracts for the long term.

The best way for these guardians to interact is the establishment of an official coordination agency prior to the bid that is solely responsible for planning and securing positive Olympic legacies, for example Rio de Janeiro's sustainability subdivision. This agency should continue to operate after the Games with the task to supervise the implementation of the sustainable legacy.

For transportation, in particular, this coordination element is vital. Crucial for Games operations are centralized planning, communication and coordination empowering *one* authority to act *across* jurisdictions (ministries, departments, planning councils, etc.). In order for transport to function like clockwork, such a coordinating entity is fundamental and should be sustained after the mega event. For example, ORTA could have taken over public transport coordination in Sydney, or ATHOC's transport division could have functioned as a link between OASA and the private transport providers in Athens. Yet, given the reduced scope of the transport task, the size of the agency would have to be adjusted to the post-Games operations.

Summary of recommendations

Cities have a trajectory of change in their urban development, which results from the working of the market system, political decision-making and other factors. The intended long-term future trajectory of change is what can be embodied in a plan. Mega events become a perturbation of this trajectory, which carry the potential either to be an opportunity for discontinuous improvement or a diversion, consumer of resources and producer of short-term activities. In the latter case, the city is afterwards left to lurch along minus whatever resources and energy it used up in achieving the Olympic goals. Maybe it returns to the trajectory it previously had; but basically, there are no long-term legacies valuable to the citizens. Alternatively, cities that have figured out how to use the currents of the Olympics to build their urban and transport infrastructure can get a big boost that yields lasting benefits. While discussed in previous chapters and in this conclusion, the following six recommendations provide bidding cities with guidelines on planning for Olympic urban and transport legacies:

Recommendation #1: Establish a hierarchy among potential Olympic legacies, for example, economic, image creation, urban regeneration. If the requirements for each deviate or impose too much of a burden on other legacies, do not bid.

Recommendation #2: Critically assess the likelihood of legacies becoming reality through the Olympics. Acknowledge potential pitfalls in executing planning, implementation and sustainment of legacies as promised in the bid.

Recommendation #3: Align planning between the event and the metropolitan strategy years in advance of submitting the bid. If a Master plan is not in place, withhold the bid.

Recommendation #4: Catalyze legacies the city actually needs. While the IOC cannot significantly deviate from its requirements (so it can ensure successful Olympics), the host cities have the flexibility to creatively adjust and leverage the requirements. In the end, it is the host city's responsibility to ensure sustainable legacies.

Recommendation #5: Identify and secure commitment from powerful guardians to protect the legacies.

Recommendation #6: Inquire about past host cities' negative legacies and ask previous hosts to point out possible pitfalls in legacy planning.

A final consideration for the IOC though: I believe the committee needs a more sophisticated screening approach for verifying candidate cities' promised legacies based on the potential host city's historic path. While the IOC has made strides in recent years to ask the right questions and implement them as evaluation criteria, the real question to be asked is whether the city is leveraging the Games as a catalyst for a sustainable, well-thought-out journey for its citizens, or merely force fitting the urban realities to Olympic dreams. The IOC has to recognize that it is a powerful agent of urban change in host cities. Furthermore, the committee has tremendous resources at hand and significant bargaining power to turn promised legacies into reality. As it seems with Rio, the IOC is asking the right questions in its bid questionnaires in regards to Olympic planning and the legacy alignment with the host city's future. Yet, Rio seems to have found a way to promise extensive legacies at the expense of large population groups. Therefore, the IOC needs a better, more independent view into the hosts and not rely solely on the bid files of the cities as those seem likely to overpromise legacies.

My findings on Olympic host cities are to a certain extent transferable to cities hosting other mega events. The impacts of the World Cup or a world exhibition might be smaller in scale, but they require similar transport preparations for the peak passenger demands, including officials, athletes and event attendees. The significance of these impacts is evident in former host cities every day. Given this tremendous influence, the number of recurring mega events, and the increasing number of new events and their growing size, there is a genuine opportunity for better-informed planners to make a difference in how cities look, and in how they function in the future by using the mega event as a tool for discontinuous change. The six Olympic cities I have discussed in this book are a good set of cases from which to draw meaningful conclusions. However, much more research is necessary to understand the broad implications mega events have for cities' urban forms and their transport systems. In this book, I focused on aligning the Olympic transport requirements with the metropolitan transport and urban development strategy of the individual cities. Further work is necessary to capture the true intention of my concept of "alignment." Beneficial regional development, new employment centers and environmental protection are likely desirable goals for future hosts. Planners have to understand and extract Olympic opportunities that can be leveraged in order to realize the broader metropolitan strategy and vision.

Using the mega event as a tool for discontinuous change requires significant public investment. The more fundamental the change being sought, the more investments will be necessary. The benefit of using the mega event as a tool is the fixed deadline of the Opening Ceremony, when all projects have to be completed and the political buy-in is achievable throughout the mega event preparation period. To ensure the Olympic vision with a great post-Games legacy becomes reality, planning has to carefully acknowledge the potential the Games bring for their urban realities and leverage them instead of focusing exclusively on the three-week transport dream.

BIBLIOGRAPHY

Acioly Jr, C. (2001) 'Reviewing urban revitalisation strategies in Rio de Janeiro: from urban project to urban management approaches', *Geoforum*, 32(4): 509–20.

Ajuntament de Barcelona (Barcelona City Government). (1988) 'Circulacio: conflicte i recepte', *Barcelona – metropolis mediterrania*, Quadern Central (8).

— (1990) *Anuari Estadistic Barcelona. Transports i circulacio* (No. ISSN 1130-4669), Barcelona: Departament d'Estadistica.

— (1993) *Anuari Estadistic de la Ciutat de Barcelona* (No. ISSN 1130-4669), Barcelona: Departament d'Estadistica.

— (1995) *Anuari Estadistic de la Ciutat de Barcelona. Transports, circulacio I communicacions* (No. ISSN 1130-4669), Barcelona: Departament d'Estadistica.

— (1999) *Anuari Estadistic de la Ciutat de Barcelona. Transports, circulacio I communicacions* (No. ISSN 1130-4669), Barcelona: Departament d'Estadistica.

Alexandridis, T. (2007) *The housing impact of the 2004 Olympic Games in Athens*, Geneva: Center on Housing Rights and Evictions (COHRE).

Andranovich, G., Burbank, M. and Heying, C. H. (2001) 'Olympic Cities: Lessons learned from Mega-Event Politics', *Journal of Urban Affairs*, 23(2): 113–31.

Anguera Argilaga, M. T. and Gomez Benito, J. (1997) *Els Jocs Olimpics Quatre Anys Despres: Avaluacio de la seva incidencia a Barcelona arran la satisfaccio dels ciutadans*, Barcelona: Universitat de Barcelona.

Anonymous. (1999) 'To traffic hell and back', *The Economist*, 351(8118): 23.

Applebome, P. (23 July 1996) 'Atlanta: Day 4 – Traffic; Athletes' Challenge: Getting There', *The New York Times*.

ARUP. (2002) *London Olympics 2012 Summary*, London: ARUP (mimeo).

Athens 2004. (1997) *Athens 2004 Candidate City. Bid Book*, Athens: Athens 2004.

Athens 2004 Transport Division. (2003) *InfoKit September 2003*, Athens: Athens 2004.

ATHOC (Athens Organizing Committee for the Olympic Games). (2002) *Athens 2004 Info Kit. Infrastructure and Transportation*, Athens: ATHOC.

— (2005) *Official Report of the XXVII Olympiad*, Athens: Liberis Publications Group.

Attiko-Metro, S.A. (1997) *Metro Development Study. Land use and socioeconomic characteristics*, Athens: Attiko Metro, S.A.

Beaty, A. (2007) *Atlanta's Olympic legacy*, Atlanta, GA: Centre on Housing Rights and Evictions (COHRE).

Ben, M., Gough, A., Masson, J., Nakhie, R., Nalder, G., Tai, S., et al. (2000) 'Transportation', in R. Freestone (ed.) *Sydney and the 2000 Olympics*, Sydney: University of New South Wales.

Bennett, S. (1999) 'Train Problems Jeopardise Hannover Expo Plans', *International Railway Journal* (November): 1–4.

Billouin, A. (1997) *Atlanta's mistakes and Sydney's resolutions for a more efficient press organization at the games of the year 2000*, Athens: The International Olympic Academy. Online. Available HTTP: http://ioa.org.gr/en/proceedings/special-sessions/cat_view/45-special-sessions-complete-books (1997 report on pp. 156–64, accessed 14 December 2011).

Black, J. (1999) 'Transport', in R. Cashman and A. Hughes (eds) *Staging the Olympics: The Event and its Impact*, Sydney: New South Wales University Press.

Blair, W., Burns, C., Giusti, C. and Sismey, K. (1996) 'Atlanta Olympics: 1996', in R. Freestone (ed.) *Urban Impacts of Olympic Games*, Sydney: University of New South Wales.

Bofill, R. (1993) 'Between Ground and Sky – remodeling the airport of Barcelona', in Ajuntament de Barcelona (ed.) *La Barcelona de 1993*, Barcelona: Ajuntament de Barcelona.

Booz•Allen & Hamilton. (1997) *Olympic Transportation System Management, System and Operations Review*, Booz•Allen & Hamilton, written for U.S. Department of Transportation, Federal Transit Administration and the Federal Highway Administration.

—— (2001) *Sydney Olympic Games Transportation Review. Final Report*, Sydney, Australia: Booz•Allen & Hamilton, written for the Athens Olympic Committee.

Borja, J. (2004) 'The city, democracy and governability: the case of Barcelona', in T. Marshall (ed.) *Transforming Barcelona. The Renewal of a European Metropolis*, New York: Routledge.

Botella, M. (1995) 'The keys to success of the Barcelona Games', in M. de Moragas and M. Botella (eds) *The Keys of success: the social, sporting, economic and communications impact of Barcelona'92*, Barcelona: Centre d'Estudis Olimpics I de l'esport (UAB).

Bovy, P. (2004a) *Mega event transport planning and traffic management – 2004*. Online. Available HTTP: www.mobility-bovy.ch/resources/33_AISTS_04.pdf (accessed 14 December 2011).

—— (2004b) *World mega-event global transport and traffic management*, Online. Available HTTP: www.mobility-bovy.ch (accessed 2 October 2008).

—— (2005) *Mega event bidding process. The role of transport*. Online. Available HTTP: http://www.mobility-bovy.ch/resources/28_CIES.bidding.05.pdf (accessed 14 December 2011).

—— (2006) 'Solving outstanding mega-event transport challenges: the Olympic experience', *Public Transport International*, 6: 32–4.

—— (2009) 'Mega-events: catalyst for more sustainable transport in Cities', paper presented at UITP Latin America mega-event and Public Transport Seminar, Recife, Brazil: November.

—— (2010) *Olympic transport and sustainability*. Online. Available HTTP: www.mobility-bovy.ch (accessed 2 May 2011).

—— (2011) *Mega-event organization and bidding: the case of transport*. Online. Available HTTP: www.mobility-bovy.ch (accessed 2 May 2011).

Bovy, P. and Liaudat, C. (2003) *Trafic de support logistique de grandes manifestations*, Lausanne: Laboratoire de mobilite et developpement territorial (mdt) de l'ecole polytechnique federale de Lausanne (EPFL).

Bovy, P., Bubka, S., Capralos, S., Elphinston, B. and Fairwheather, K. (2004) *Report by the IOC Candidature acceptance working group to the IOC executive board*, Lausanne: IOC.

Brandão, Z. (2006) 'Urban Planning in Rio de Janeiro: a Critical Review of the Urban Design Practice in the Twentieth Century', *City & Time*, 2(2): 37–50.

Brunet, F. (1993) *Economy of the 1992 Barcelona Olympic Games*, Lausanne: IOC Olympic Study and Research Centre.

Bulkin, R. (1995) *Official guide to Atlanta and the Olympic Summer Games*, Atlanta, GA: Frommer.

Burbank, M., Andranovich, G. and Heying, C. H. (2001) *Olympic dreams: the impact of mega-events on local politics*, Boulder, CO: Lynne Rienner.

Capel, H. (2007) *El debate sobre la construccion de la ciudad y el llamado 'modelo Barcelona'*, Barcelona: Universidad di Barcelona. Online. Available HTTP: www.ub.es/geocrit/sn/sn-233.htm (accessed 15 February 2009).

Cashman, R. (2002) *Impact of the Games on Olympic host cities: university lecture on the Olympics*, Barcelona: Centre d'Estudis Olimpics (UAB), International Chair in Olympism (IOC-UAB). Online. Available HTTP: http://olympicstudies.uab.es/lectures/web/pdf/cashman.pdf (accessed 5 November 2008).

— (2006) *The Bitter-Sweet Awakening: the Legacy of the Sydney 2000 Olympic Games*, Sydney: Walla Walla Press.

— (2011) *Sydney Olympic Park 2000–2010*, Sydney: Walla Walla Press.

Cashman, R. and Hughes, A. (eds) (1999) *Staging the Olympics: The Event and its Impact*, Sydney: University of New South Wales Press.

Chalkley, B. and Essex, S. (1999) 'Urban Development through hosting international events: a history of the Olympic Games', *Planning Perspectives*, 14: 369–94.

Chapman, J. (2000) 'Impact of building roads to everywhere', in R. D. Bullard, A. O. Torres and G. S. Johnson (eds) *Sprawl city: race, politics, and planning in Atlanta*, Washington, D.C.: Island Press.

Chorianopoulos, I., Pagonis, T., Koukoulas, S. and Drymoniti, S. (2010) 'Planning, competitiveness and sprawl in the Mediterranean city: the case of Athens', *Cities*, 27(4): 249–59.

Christy, D., Erken, J., McCarry, N., McKay, K. and O'Dwyer, L. (1996) 'Major Events', in R. Freestone (ed.) *Urban Impacts of Olympic Games*, Sydney: Working Paper 96/1, School of Planning and Urban Development, University of New South Wales.

CIO (Center International Olympique). (1985) *Request to the International Olympic Committee by the Barcelona Candidature for the hosting of the Games of the XXVth Olimpiad*, Lausanne: IOC.

Clark, G. (2008) *Local development benefits from staging global events*, Paris: OECD.

Clean Air 2000 Project Team. (1996) *Shaping Sydney's Transport – a framework for reform*, Sydney: NRMA limited.

CMB (Corporacion Metropolitana de Barcelona). (1976) *Programa de actuacion*, Barcelona: Corporacion Metropolitana de Barcelona.

Coaffee, J. (2007) 'Urban Regeneration and Renewal', in J. R. Gold and M. M. Gold (eds) *Olympic Cities City Agendas, Planning and the World's Games, 1896–2012*, New York: Routledge.

Cochrane, A., Peck, J. and Tickell, A. (1996) 'Manchester Plays Games: Exploring the Local Politics of Globalisation', *Urban Studies*, 33(8): 1319–36.

CODA (Corporation for Olympic Development in Atlanta). (1993) *Master Olympic development program for City of Atlanta*, Atlanta, GA: BellSouth Telecommunications.

Connor, J., Heretis, F., Liu, S., Press, S., Stefan, D., Watson, C., et al. (2000) 'Planning Initiatives', in R. Freestone (ed.) *Sydney and the 2000 Olympics*, Sydney: University of New South Wales.

Cross, B. (1992) 'Barcelona prepares for the Olympics', *UTI*, March/April: 12–15.

Culat, E. (8 July 1992) 'El ayuntamiento edita 1,3 milliones de guias sobre movilidad olimpica', *El observador*.

Curnow, A. (2000) 'Environmental issues drive transport plans', *Stadia*, August: 62–8.

Currie, G. (1998) 'The Planning and Performance of Mass Transit Operating Strategies for Major Events – The Atlanta Olympics and the 1996 Melbourne Formula 1 Grand Prix', paper presented at Australian Institute of Transport Planning and Management. *National Conference*. "Transport Planning for Major Events", Melbourne.

—— (2007) *Olympic Transport Planning – Lessons for London*, Institute of Transport Studies, Monash University, Melbourne.

Department of Planning. (1968) *Sydney Region: outline plan, 1970–2000*, Sydney: New South Wales Government.

—— (1988) *Sydney into its third century*, Sydney: New South Wales Government.

—— (1992) *Updating the metropolitan strategy*, Sydney: New South Wales Government.

—— (1993) *Sydney's future: a discussion paper on planning the greater metropolitan region*, Sydney: New South Wales Government.

—— (1995) *Cities for the 21st century – Integrated urban management for Sydney, Newcastle, the Central Coast, and Wollongong*, Sydney: New South Wales Government.

Department of Transport. (1995) *Integrated Transport Strategy for the Greater Metropolitan Region*, Sydney: New South Wales Government.

Dimitriou, D., Karlaftis, M., Kepaptsoglou, K. and Stathopoulos, A. (2004) 'Public Transportation during the Athens 2004 Olympics: Facts, Performance and Evaluation', in P. A. Stathopoulos (ed.) *Athens 2004 Summer Olympics: A Compendium of Best Transportation Practices*, Athens: Hellenic Institute of Transportation Engineers.

Dunn, K. M. and McGurick, P. M. (1999) 'Hallmark Events', in R. Cashman and A. Hughes (eds) *Staging the Olympics: The Event and its Impact*, Sydney: New South Wales University Press.

ECMT (European Conference of Ministers of Transport). (2003) *Transport and Exceptional Public Events* (Vol. Roundtable 122), Paris: OECD Publications Service.

Eitzen, D. S. (1996) 'Classism in sport: the powerless bear the burden', *Journal of Sport and Social Issues*, 20(1): 95–105.

Essex, S. and Chalkley, B. (2002) 'The Infrastructural Legacy of the Summer and Winter Olympic Games – A comparative Analysis', *The Legacy of the Olympic Games: 1984–2000*, Lausanne: IOC.

—— (2004) 'Mega-sporting events in urban and regional policy: a history of the Winter Olympics', *Planning Perspectives*, 19(2): 201–32.

—— (2005) *The Olympic Games: catalyst of urban change*. Online. Available HTTP: www.otromadrid.org/m2012/leer.php?id=383 (accessed 22 June 2008).

—— (2008) *Research Themes. The Olympic Games: catalyst of urban change*. Online. Available HTTP: www.plymouth.ac.uk/files/extranet/docs/SSB/eurolympicgames.pdf (accessed 14 December 2011).

Esteban, J. (1999) *El proyecto urbanístico. Valorar la periferia y recuperar el centro*, Barcelona: Universitat de Barcelona.

FIFA (Federation Internationale de Football Association). (2007) *Brazil bid. FIFA inspection report for the 2014 FIFA World Cup*, Geneva: FIFA.

French, S. P. and Disher, M. E. (1997) 'Atlanta and the Olympics. A One-Year Retrospective', *Journal of the American Planning Association*, 63(3): 379–92.

Gaffney, C. (2010) 'Mega-events and socio-spatial dynamics in Rio de Janeiro, 1919–2016', *Journal of Latin American Geography*, 9(1): 7–29.

— (2011a) *The Good News*. Online. Available HTTP: www.geostadia.com (accessed 2 September 2011).

— (2011b) *Jogos Mundiais Militares: legado do Pan-Americano ou legado de exclusão? (Military World Games, Pan American legacy or legacy of exclusion?)*. Online. Available HTTP: http://observatoriodasmetropoles.net/index.php?option=com_content&view=articl e&id=1719&catid=45&Itemid=88&lang=pt (accessed 2 September 2011).

— (2011c) *Laws, Speculation, Eminent Domain, Stadia*. Online. Available HTTP: www. geostadia.com (accessed 2 September 2011).

Garlick, R. (2009) 'The Game Plan', *Regeneration & Renewal*, 16 February: 14–15.

Gee, J., Hay, A. and Bell, S. (1996) *Public Transport Travel Patterns in the Greater Sydney Metropolitan area 1981 to 1991*, Sydney: Transport Data Centre – New South Wales Department of Transport.

Getz, D. (1999) 'The Impacts of Mega Events on Tourism: Strategies for Destinations', in T. D. Andersson, C. Persson, B. Stahlberg and L.-I. Ström (eds) *The Impacts of Mega Events*, Ornsköldsvik: European Tourism Research Institute (ETOUR).

Gibbons, R. (1983) 'The fall of the giant: trams versus trains and buses in Sydney 1900– 1961', in G. Witherspoon (ed.) *Sydney's transport*, Sydney: Hale & Iremonger Pty Limited.

Gold, J. R. and Gold, M. M. (eds) (2007) *Olympic Cities – City Agendas, Planning, and the World's Games, 1896–2012*, London: Routledge.

— (eds) (2010) *Olympic Cities – City Agendas, Planning, and the World's Games, 1896–2016* (2nd ed.), London and New York: Routledge.

Goldberg, D. (6 July 1996) 'Will the lessons we learned be more than transitory?', *The Atlanta Journal Constitution*.

Goldberg, D. and Hill, A. E. (30 July 1996) 'Commuters, spectators return to the roads', *The Atlanta Journal Constitution*.

Gratton, C. and Preuss, H. (2008) 'Maximizing Olympic impacts by building up legacies', *International Journal of the History of Sport*, 25(14): 1922–38.

Greater London Authority. (2004) *The London Plan – Spatial Development Strategy for Greater London*, London: Greater London Authority.

— (2007) *2012 Summer Olympic Games London England London 2012 sustainability plan – Annual Report 2007*, London: Greater London Authority.

— (2011) *The London Plan – Spatial Development Strategy for Greater London*, London: Greater London Authority.

Green, S. (19 July 2011) 'Rio-SP High-Speed Train Late for Olympics', *The Rio Times*. Online. Available HTTP: http://riotimesonline.com/brazil-news/front-page/rio-sp-high-speed-train-late-for-olympics/ (accessed 6 December 2011).

Guàrdia i Bassols, M., Monclús, F. J., Oyón, J. L. and Centre de Cultura Contemporània de Barcelona. (1994) *Atlas histórico de ciudades europeas*, Barcelona: Salvat.

Hall, C. M. (1989) 'The Definition and Analysis of Hallmark Tourist Events', *GeoJournal*, 19(3): 263–8.

— (1996) 'Hallmark events and urban reimaging strategies: Coercion, community, and the Sydney 2000 Olympics', in L. C. Harrison and W. Husbands (eds) *Practicing responsible tourism: International case studies in tourism planning, policy, and development*, New York: John Wiley.

Harvey, D. (1989) *The Urban Experience*, Oxford: Blackwell.

Helling, A. (1997) *Changing intra-metropolitan accessibility in the U.S. – evidence from Atlanta*, Exeter: Pergamon.

Hensher, D. and Brewer, A. (2002) 'Going for gold at the Sydney olympics: How did transport perform?', *Transport Reviews*, 22(4): 381–99.

Hersh, P. (29 May 2008) 'Transportation improvements key to Chicago's Olympic bid', *Chicago Tribune*. Online. Available HTTP: http://featuresblogs.chicagotribune.com/theskyline/2008/05/transportation.html (accessed 3 September 2011).

Hill, A. E. (23 July 1996) 'Planning pays off as downtown remains snarl-free', *The Atlanta Journal Constitution*.

— (29 July 1996) 'Gridlock expected if commuters drive', *The Atlanta Journal Constitution*.

Hiller, H. H. (2006) 'Post-event Outcomes and the Post-modern Turn: The Olympics and Urban Transformations', *European Sports Management Quaterly*, 6(4): 317–32.

Horne, J. (2007) 'The Four "Knowns" of Sports Mega-Events', *Leisure Studies*, 26(1): 81–96.

Horne, J. and Manzenreiter, W. (2006) *Sports mega-events: social scientific analyses of a global phenomenon*, Malden, MA: Wiley-Blackwell.

House of Commons. (2003) *A London Olympic Bid for 2012*, London: The Stationary Office Limited.

IOC (International Olympic Committee). (1992) *Manual for cities bidding for the Olympic Games*, Lausanne: IOC.

— (1997) *Report of the IOC Evaluation Commission for the Games of the XXVIII Olympiad in 2004*, Lausanne: IOC.

— (2000) *Report by the IOC candidature acceptance working group to the executive board of the International Olympic Committee. Candidature acceptance procedure of the Games of the XXIX Olympiad 2008*, Lausanne: IOC.

— (2003d) *Olympic Charter in force as from 4 July 2003*, Lausanne: IOC.

— (2004a) *2012 Candidature Procedure and Questionnaire: Games of the XXX Olympiad in 2012*, Lausanne: IOC.

— (2004b) *Report by the IOC Candidature Acceptance Working Group to the IOC executive board*, Lausanne: IOC.

— (2005) *Report of the IOC Evaluation Commission for the Games of the XXX Olympiad in 2012*, Lausanne: IOC.

— (2007) *Olympic Charter in force as from 7 July 2007*, Lausanne: IOC.

— (2008a) *2016 Candidature Procedure and Questionnaire: Games of the XXXI Olympiad in 2016*, Lausanne: IOC.

— (2008b) *Games of the XXXI Olympiad 2016 Working group report*, Lausanne: IOC.

— (2009) *Report of the 2016 IOC evaluation commission. Games of the XXXI Olympiad*, Lausanne: IOC.

— (2011) *Olympic Charter in force as from 8 July 2011*, Lausanne: IOC.

IplanRio. (1996) *Rio Cidade: o Urbanismo de Volta as Ruas*, Rio de Janeiro: Mauad.

Jacana Consulting Pty Ltd. (1996) *Transport Infrastructure for Homebush Bay and Sydney Olympics 2000*, Sydney: Green Games Watch 2000 Inc.

Jago, L., Dwyer, L., Lipman, G. and Vorster, S. (2010) 'Optimising the potential of mega-events: an overview', *International Journal of Event and Festival Management*, 1(3): 220–37.

James, P. and Myer, A. (1997) *Sydney 2000 – Strategy for a Car-Free Olympic Games*, Sydney: Green Games Watch 2000 Inc.

Kassens-Noor, E. (2010a) 'Peak Demands in Transport Systems – an Agenda for Research'. Paper ID G6-01541, in J. Viegas and R. Macário (eds), *Selected Proceedings of the 12th World Conference on Transport Research Society*, Lisbon, Portugal, 11–15 July.

— (2010b) 'Sustaining the Momentum – the Olympics as potential Catalysts for enhancing Urban Transport', *Transportation Research Record: Journal of the Transportation Research Board*, 2187: 106–13.

Kissoudi, P. (2010) 'Athens' Post-Olympic Aspirations and the Extent of their Realization', *The International Journal of the History of Sport*, 27(16–18): 2780–97.

Klein, F. C. (10 August 1992) 'Barcelona bids Farewell as Games Cross finish line', *The Wall Street Journal*.

Landscape Online. (2011) *London Olympic Design By EDAW*. Online. Available HTTP: www.landscapeonline.com/research/article/5407 (accessed 8 August 2001).

Lenskyj, H. (2002) *The best Olympics ever?: social impacts of Sydney 2000*, Albany: State University of New York Press.

Leonardsen, D. (2007) 'Planning of Mega Events: Experiences and Lessons', *Planning Theory & Practice*, 8(1): 11–30.

LOCOG (London Organizing Committee for the Olympic Games). (2005) *London 2012 Candidacy File. Bid book*, London: LOCOG.

London 2012 Ltd. (2004) *Response to the questionnaire for cities applying to become Candidate cities to host the Games of the XXX Olympiad and the Paralympic Games in 2012*, London: London 2012.

Lopes, R. (2003) *O Planejamento Estratégico da Cidade do Rio de Janeiro – Um Processo de Transformação*. Online. Available HTTP: www.iets.org.br/article.php3?id_article=623 (accessed 10 August 2011).

Malfas, M., Theodoraki, E. and Houlihan, B. (2004) 'Impacts of the Olympic Games as mega-events', *Proceedings of the Institute of Civil Engineers, Municipal Engineer*, 157(ME3): 209–20.

Maricato, E. (2002) *Brasil, Cidades – Alternativas para a crise urbana*, Rio de Janeiro: Ed. Vozes.

MARTA (Metropolitan Atlanta Rapid Transit Authority). (2009) *History of MARTA – 1990–1999*. Online. Available HTTP: http://www.itsmarta.com/marta-past-and-future.aspx (accessed 14 December 2011).

Martorell, J., Bohigas, O., Makcay, D. and Puigdomènech, A. (1991) *La Villa Olimpica Barcelona 92. Arquitectura. Parques. Puerto deportivo*, Barcelona: Holding Olimpico SA.

Mateu Cromo. (1996) *Plano de Barcelona*, Madrid: S.A. Pinto.

MEPPW (Ministry of Environment Planning and Public Works). (1985) *The Master plan of Athens*, Athens: Ministry of Environment Planning and Public Works.

Milakis, D., Vlastos, T. and Barbopoulos, N. (2008) 'Relationships between Urban Form and Travel Behaviour in Athens, Greece. A Comparison with Western European and North American Results', *EJTIR*, 8(3): 201–15.

Millet, L. (1995) 'The Games of the City', in M. de Moragas and M. Botella (eds) *The keys of success: the social, sporting, economic, and communications impact of Barcelona '92*, Barcelona: Centre d'Estudis Olimpics I de l'esport (UAB).

Minis, I., Angelopoulos, J. and Kyrioglou, G. (2009) 'Car fleet planning and management models for large event transport: the Athens 2004 Olympic Games', *Transportation Planning & Technology*, 32(2): 135–61.

Minis, I. and Tsamboulas, D. A. (2008) 'Contingency Planning and War Gaming for the Transport Operations of the Athens 2004 Olympic Games', *Transport Reviews*, 28(2): 259–80.

Ministry of the Environment. (2009) *Athens history*. Online. Available HTTP: www.minenv.gr/3/31/313/31303/e3130303.html (accessed 18 August 2010).

MLR (Metro Light Rail). (2008) *Technical Statistics, Metro Light Rail*. Online. Available HTTP: www.metrotransport.com.au/network/overview/lightrail (accessed 15 August 2009).

Monroe, D. (29 July 1996) 'Cars, delays finally hit downtown', *The Atlanta Journal Constitution*.

Mosby, A. (1992) *The year of Spain*, Lausanne: IOC.

Munoz, F. (1997) 'Historic Evolution and Urban Planning Typology of Olympic Villages', in M. De Moragas, M. Llines and B. Kidd (eds) *Olympic Villages: a hundred Years of Urban Planning and Shared Experiences*, Lausanne: International Olympic Museum.

NBCSports. (1991) *How Barcelona got the Games*, Barcelona, Spain.

Nelson, A. C. (1993) *The role of regional development management in central city revitalization: case studies and comparisons of development patterns in Atlanta, Georgia, and Portland, Oregon*, New Orleans, LA: University of New Orleans.

OASA (Οργανισμός Αστικών Συγκοινωνιών Αθηνών) (2009) *Athens Urban Transport Organization – History*. Online. Available HTTP: www.oasa.gr/content.php?id=istoria (in Greek, accessed 14 December 2011).

OCA (Olympic Coordination Authority). (1995a) *Homebush Bay Development Guidelines (Vol 4): Transport Strategy*, Sydney: Olympic Coordination Authority.

ODA (Olympic Delivery Authority). (2009) *Transport Plan for the London 2012 Olympic and Paralympic Games – second edition consultation draft*, London: Olympic Delivery Authority.

— (2010) *Transport update. Issue 7: Sustainable Transport*, London: Olympic Delivery Authority.

OECD (Organization for Economic Co-operation and Development). (2004) *OECD territorial reviews, Athens Greece*, Paris: OECD Publication Service.

Olympic Movement. (2008) *Structure of the Olympic Movement > International Olympic Committee (IOC)*. Online. Available HTTP: www.olympic.org/uk/utilities/faq_detail_uk.asp?rdo_cat=12_24_0&faq=146 (accessed 9 November 2008).

ORTA (Olympic Roads and Transit Authority). (2001) *Nothing bigger than this – Transport for the Sydney 2000 Olympic and Paralymic Games*, Sydney: Olympic Roads and Transport Authority.

Ostertag, M. (2010) 'No transport white elephants – Interview with Prof. P. Bovy', *ITS Magazine – The Magazine for Intelligent Traffic Systems*, 2: 16–8.

Owen, K. A. (2002) 'The Sydney 2000 Olympics and Urban Entrepreneurialism: Local Variations in Urban Governance', *Australian Geographical Studies*, 40(3): 323–36.

Palese, B., Millais, C., Posner, R., Koza, F., Mealey, E., McLaren, W., et al. (2000) *How green the Games? Greenpeace's environmental assessment of the Sydney 2000 Olympics*, Sydney: Greenpeace International & Greenpeace Australia Pacific.

Pappas, E. (2003) *Will 2004 Olympics Destroy Ancient Greek Battleground?* Online. Available HTTP: http://news.nationalgeographic.co.in/news/2003/08/0801_030801_marathonathens.html (accessed 1 August 2010).

PHF (Pan-Hellenic Federation of University Graduate Engineer Civil Servant Unions). (2001) *Olympiaka Erga*. Online. Available HTTP: www.emdydas.gr/2004_porisma.htm (accessed 2 February 2009).

Pires, H. F. (2010) *Planejamento e intervenções urbanísticas no Rio de Janeiro: a utopia do plano estratégico e sua inspiração catalã*. Online. Available HTTP: www.ub.es/geocrit/b3w-895/b3w-895-13.htm (accessed 31 July 2011).

Plano Estratégico da Cidade do Rio de Janeiro. (1996) *Plano Estratégico da Cidade do Rio de Janeiro, Rio Sempre Rio*, Rio de Janeiro: Imprensa da cidade.

Powell, D. (23 October 2007) 'Atlanta took gold in the Chaos Games'. Online. Available HTTP: www.timesonline.co.uk/tol/sport/olympics/london_2012/article2719542.ece (accessed 5 April 2009).

Poynter, G. (2004) *From Beijing to bow bells: Measuring the Olympics effect*, London: University of East London, London East Research Institute.

Poynter, G. and MacRury, I. (eds) (2009) *Olympic Cities: 2012 and the Remaking of London*, Farnham: Ashgate.

Prefeitura do Rio de Janeiro. (2010) *Pos 2016 o Rio mais integrado e competitivo, plano estratégico 2009–2012*, Rio de Janeiro: Prefeitura do rio de janeiro.

Preuss, H. (2004) *The Economics of Staging the Olympics. A Comparison of the Games 1972–2008*, Cheltenham: Edward Elgar.

Psaraki, V. (1994) *Priorisation measures in Greece*, Athens: OASA.

Punter, J. (2005) 'Urban design in central Sydney 1945–2002: Laissez-Faire and discretionary traditions in the accidental city', *Progress in Planning*, 63(1): 11–160.

Pyo, S., Cook, R. and Howell, R. L. (1988) 'Summer Olympic tourist market – learning from the past', *Tourism Management*, 9(2): 137–44.

Riera, P. (1993) *Rentabilidad social de las infraestructuras: las Rondas de Barcelona. un analisis coste-beneficio*, Madrid: Editorial Civitas S.A.

Rio 2016. (2009) *Candidature file for Rio de Janeiro to host the 2016 Olympic and Paralympic Games*, Rio de Janeiro: Rio 2016.

Ritchie, J. B. R. (1984) 'Assessing the Impact of Hallmark Events: Conceptual and Research Issues', *Journal of Travel Research*, 23(1): 2–11.

Robbins, D., Dickinson, J. and Calver, S. (2007) 'Planning Transport for Special Events: A Conceptual Framework and Future Agenda for Research', *International Journal of Tourism Research*, 9(5): 303–14.

Roche, M. (2006) 'Mega-events and modernity revisited: globalization and the case of the Olympics', in J. Horne and W. Manzenreiter (eds) *Sports Mega-Events: Social Scientific Analyses of a Global Phenomenon*, Oxford: Blackwell.

Rubalcaba-Bermejo, L. and Cuadrado-Roura, J. R. (1995) 'Urban Hierarchies and Territorial Competition in Europe: Exploring the Role of Fairs and Exhibitions', *Urban Studies*, 32(2): 379–400.

Rushin, S. (1996) 'Those buses were a bust – ACOG's transportation fiasco', *Sports Illustrated*, (12 August).

Ruthheiser, D. (2000) *Imagineering Atlanta: the politics of place in the city of dreams*, New York: Verso.

Sànchez, A. B., Plandiura, R. and Valiño, V. (2007) *Barcelona 1992: international events and housing rights: a focus on the Olympic Games*, Geneva: Center on Housing Rights and Evictions (COHRE).

Schwandl, R. (2004) *Barcelona Metro History*. Online. Available HTTP: www.urbanrail.net/eu/es/bcn/bcn-metro-history.htm (accessed 14 December 2011).

Searle, G. (2002) 'Uncertain Legacy: Sydney's Olympic Stadiums', *European Planning Studies*, 10(7): 846–60.

Senn, A. (1999) *Power, Politics and the Olympic Games*, Champaign, IL: Human Kinetics.

Servicios Tecnicos de la CMB. (1993) 'Una red viaria definitiva', in Ajuntament de Barcelona (ed.) *La Barcelona de 1993*, Barcelona: Ajuntament de Barcelona.

Short, J. R. (2004) *Global metropolitan: globalizing cities in a capitalist world*, London: Routledge.

— (2008) 'Globalization, cities and the Summer Olympics', *City: analysis of urban trends, culture, theory, policy, action*, 12(3): 321–40.

SOCOG (Sydney Organizing Committee for the Olympic Games). (2001) 'Preparing for the Games', *Evaluation Report on the Sydney 2000 Olympics*, Sydney: IOC.

Sydney 2000. (1993) *Sydney 2000 Candidate City. Bid Book*, Sydney: Sydney 2000.

Syme, S., Shaw, B., Fenton, D. and Mueller, W. (1989) *The Planning and Evaluation of Hallmark Events*, Aldershot: Ashgate.

Telloglou, T. (2004) *The City of the Games*, Athens: Hestia Publishers & Booksellers (in Greek).

TMB (Transports Metropolitans de Barcelona). (2008) *What is TMB?*. Online. Available HTTP: www.polis-online.org/fileadmin/hot_topic/Barcelona_Conference/Presentations/Public_Transport_Control_Centre.pdf (accessed 28 February 2009).

U.S. Department of Transportation. (1997) *1996 Atlanta Centennial Olympic Games and Paralympic Games Event Study*. Online. Available HTTP: http://trid.trb.org/view.aspx?id=496229 (accessed 14 December 2011).

Vaeth, E. (1996) 'Olympics: A defining moment in Atlanta's history. Was it worth it?', *Atlanta Business Chronicle*. Online. Available HTTP: http://atlanta.bizjournals.com/atlanta/stories/1998/06/15/focus17.html (accessed 30 November 2008).

Waghorn, D. (2000) 'Past plans', in R. Freestone (ed.) *Sydney and the 2000 Olympics*, Sydney: University of New South Wales.

Weirick, J. (1999) 'Urban Design', in R. Cashman and A. Hughes (eds) *Staging the Olympics: The Event and its Impact*, Sydney: New South Wales University Press.

Whitten, C. (2000) 'Sydney's bid', in R. Freestone (ed.) *Sydney and the 2000 Olympics*, Sydney: University of New South Wales.

WWF-UK. (2004) *No gold medal for the environment in the Athens Olympics*. Online. Available HTTP: www.wwf.org.uk/wwf_articles.cfm?unewsid=1277 (accessed 14 December 2011).

Yarbrough, C. R. (2000) *And They Call Them Games: An Inside View of the 1996 Olympics*, Atlanta, GA: Mercer University Press.

Zagorianakos, E. (2004) 'Athens 2004 Olympic Games' Transportation Plan: a missed opportunity for Strategic Environmental Assessment (SEA) integration?', *Journal of Transport Geography*, 12(2): 115–25.

UNPUBLISHED SOURCES

★ *Note: these documents were made available to the author and publication was allowed by permission of the International Olympic Committee (IOC); regular 25-year embargo lifted by the IOC President J. Rogge.*
★★ *Note: these documents are only accessible in person; no loans.*

Acebillo Marin, J. A. (1992) *De la Plaza Trilla a la Villa Olimpica.* Obtained from Centre for Olympic Studies (CEO) at the Universitat Autònoma de Barcelona (Barcelona, Spain). Research archives: CEO / transport general.★★

ACOG (Atlanta Organizing Committee for the Olympic Games). (1992) *Master Plan Summary.* Obtained from Olympic Museum (Lausanne, Switzerland). Olympic Studies Centre research archives: CIO / Summer Olympic Games of Atlanta 1996 – transport.★

— (1993a) *Official Report of the Atlanta Committee for the Olympic Games.* Obtained from Olympic Museum (Lausanne, Switzerland).★★

— (1993b) *Olympic Transportation Systems Study.* Obtained from Atlanta History Center (Atlanta, USA). Research archives: 26th 96 Atl., GA, Subject File Box 6, Transportation, Kenan Research Centre, Atlanta History Center Archives.★★

— (1993c) *Press Information Guide.* Obtained from Olympic Museum (Lausanne, Switzerland).

— (1994) *Olympic Transportation System Plan for the 1996 Centennial Olympic Games.* Obtained from Olympic Museum (Lausanne, Switzerland). Olympic Studies Centre research archives: CIO / Summer Olympic Games of Atlanta 1996 – transport.★

— (1995a) *Dossier for the chef de mission.* Obtained from Australian Centre for Olympic Studies (ACOS) at the University of Technology Sydney (UTS). Olympic and event studies collection.★★

— (1995b) *Report to the XIth general assembly of the Association of Summer Olympic International Federations.* Obtained from Australian Centre for Olympic Studies (ACOS) at the University of Technology Sydney (UTS). Olympic and event studies collection.★★

Ajuntament de Barcelona (Barcelona City Government). (1989a) *Conveni d'installacio del sistema de gestio del transit a la xarxa d'accesses i cinturons de Barcelona.* Obtained from

Olympic Museum (Lausanne, Switzerland). Olympic Studies Centre research archives: CIO / Summer Olympic Games of Barcelona 1992 – transport.★

— (1989b) *Organitzacio de la gestio del transit i del transport a l'area metropolitana de Barcelona.* Obtained from Olympic Museum (Lausanne, Switzerland). Olympic Studies Centre research archives: CIO / Summer Olympic Games of Barcelona 1992 – transport.★

Alevras, N. (2003) *Fax regarding metro trains to the airport.* Obtained from IOC Archives / Summer Olympic Games of Athens 2004 – transport.★

Allachin, C. (Urban Designer and Architect working for Six Degrees Pty Ltd.) (2008) *Interview:* 'The legacy of Homebush Bay – the good and the bad', Sydney: 22 February 2008.

ARC (Atlanta Regional Commission). (1996) *Transportation planning activities for the centennial Olympic Games: after action report.* Obtained from Atlanta History Center (Atlanta, USA). Research archives: 26th 96 Atl., GA, Subject File Box 6, Transportation, Kenan Research Centre, Atlanta History Center Archives.★★

— (1999) *Atlanta regional transportation plan: 2010.* Obtained from Kenan Research Centre, Atlanta History Center Archives.★★

ASOIF (Association of Summer Olympic International Federations). (1990) *Questionnaire for Candidate Cities for the organization of the 1992 Olympic Games.* Obtained from Olympic Museum (Lausanne, Switzerland). Olympic Studies Centre research archives: CIO / Summer Olympic Games of Barcelona 1992 – transport.★

ATHOC (Athens Organizing Committee for the Olympic Games). (2001) *Olympic Transport Strategic Plan.* Obtained from J. Frantzeskakis.★★

— (2004) *Olympic Transport Plan – transport operations for the ATHENS 2004 Olympic Games.* Obtained from IOC Archives / Summer Olympic Games of Athens 2004 – transport.★

Attiki-Odos. (2007) *When the result exceeds expectations.* Obtained from C. Halkias.★★

Attiko-Metro, A. E. (1996a) *Athens Metro Fact Book.* Obtained from IOC Archives / Summer Olympic Games of Athens 2004 – transport.★

— (1996b) *Metro Development Study – presentation from Attiko Metro A.E.* Obtained from K. Deloukas.★★

Babis, H. (Head of Traffic Management, ATHOC). (2007) *Interview:* 'Traffic Management Center', Athens: 31 August 2007.

Barreiro, F., Costa, J. and Vilanova, J. (1993) *Impactos urbanisticos, economicos y sociales de los Juegos Olimpicos de Barcelona '92*, Barcelona: CIREM (Centre d'Iniciatives I Recerques europees a la mediterrania).★

Black, J. (Foundation Professor of Transport Engineering at the University of New South Wales, Advisor to the NSW Government). (2008) *Interview:* 'Preparing Sydney for the transport challenge to host the Olympic Games', Sydney: 14 February 2008.

Boitelle, R., Novikov, I. and Schreiber, H. (1990) *Report of the ASOIF delegation for Atlanta. Candidate Cities for the organization of the 1996 Olympic Games.* Obtained from Olympic Museum (Lausanne, Switzerland). Olympic Studies Centre research archives: CIO / Summer Olympic Games of Atlanta 1996 – transport.★

Bovy, P. (2000) *Transport. Sydney 2000 Olympics – Observation report.* Obtained from Olympic Museum (Lausanne, Switzerland). Olympic Studies Centre research archives: CIO / Summer Olympic Games of Sydney 2000 – transport.★

Bovy, P. and Protopsaltis, P. 'Games benefits and legacy – Sustainable Development as it relates to the Games', paper presented at private seminar 2012 Applicant City, Lausanne, Switzerland: 8 October 2003.

Brockhoff, J. (Policy Manager, Metropolitan Strategy, Department of Planning, NSW). (2008) *Interview:* 'Legacies of the Olympic Games', Sydney: 21 February 2008.

Brunet, F. (Professor of Economics at Universitat Autonoma de Barcelona). (2007) *Interview:* 'The Olympics in Barcelona and their long-term impacts', Barcelona: 12 June 2007.

Cartalis, C. 'Games benefits and legacy – Sustainable Development as it relates to the Games', paper presented at Private seminar 2012 Applicant City, Lausanne: 9 October 2003.

CEAM (centro de estudios y asesoramiento metalurgico). (1987) *L'impacte industrial dels JJ.OO. Barcelona '92.* Obtained from Centre for Olympic Studies (CEO) at the Universitat Autònoma de Barcelona (Barcelona, Spain). Research archives: CEO / transport general.★★

Chalkias, B. (Managing Director Attiki Odos). (2007) *Interview:* 'The Attiki Odos', Athens, Greece: 31 August 2007.

CMB (Corporacion Metropolitana de Barcelona). (1986) *La Movilidad obligada en el area metropoltana de Barcelona.* Obtained from Centre for Olympic Studies (CEO) at the Universitat Autònoma de Barcelona (Barcelona, Spain). Research archives: CEO / transport general.★★

COOB'92 (Comite Organizador Olimpico de Barcelona '92). (1990) *Official Presentation Barcelona to the International Olympic Committee.* Obtained from Olympic Museum (Lausanne, Switzerland). Olympic Studies Centre research archives: CIO / Summer Olympic Games of Barcelona 1992 – transport.★

Currie, G. (Professor of Transportation Planning at Monash University). (2008) *Interview:* 'Olympic Transport Performance', Melbourne: 22 February 2008.

Dahlberg, B. (1996) *Business as usual – Strategies for the Summer of 1996.* Obtained from 26th 96 Atl., GA, Subject File Box 1, Kenan Research Centre, Atlanta History Centre Archives.★★

Daniels, R. (Manager Metropolitan Planning, Department of Planning, NSW). (2008) *Interview:* 'Transport planning for Homebush and beyond', Sydney: 21 February 2008.

De Frantz, A., McLatchey, C., Bovy, P. and Dubi, C. (1999) *Third Impression report on Sydney 2000 Olympic transportation.* Obtained from Olympic Museum (Lausanne, Switzerland). Olympic Studies Centre research archives: CIO / Summer Olympic Games of Sydney 2000- transport.★

De Frantz, A., Montgomery, P., Bovy, P. and Dubi, C. (1998) *Second Impression report on Sydney 2000 Olympic transportation. Mission to Sydney during the first royal easter show at Homebush Bay.* Obtained from Olympic Museum (Lausanne, Switzerland). Olympic Studies Centre research archives: CIO / Summer Olympic Games of Sydney 2000 – transport.★

De Frantz, A., Palmer, R., Bovy, P. and Dubi, C. (1997) *IOC Transportation working group. Sydney 2000 Olympic Games. First impression report.* Obtained from Olympic Museum (Lausanne, Switzerland). Olympic Studies Centre research archives: CIO / Summer Olympic Games of Sydney 2000 – transport.★

— (1999) *Fourth impression report on Sydney 2000 Olympic transportation.* Obtained from Olympic Museum (Lausanne, Switzerland). Olympic Studies Centre research archives: CIO / Summer Olympic Games of Sydney 2000 – transport.★

De Frantz, A., Palmer, R., Bovy, P. and Toulson, S. (2000) *Fifth impression report on Sydney 2000 Olympic transportation.* Obtained from Olympic Museum (Lausanne, Switzerland). Olympic Studies Centre research archives: CIO / Summer Olympic Games of Sydney 2000 – transport.★

Delegacion de la 5a zona. (1989) *Cronologia de la remodelacio de la xarca ferroviaria de Barcelona.* Obtained from Centre for Olympic Studies (CEO) at the Universitat Autònoma de Barcelona (Barcelona, Spain). Research archives: CEO / transport general.★★

Deloukas, A. (Business planning manager, Attiko Metro S.A.). (2007) *Interview:* 'Expansion of the metro line and sustainable transport', Athens: 3 September 2007.

Department of Urban Affairs and Planning. (1997) *A framework for growth and change – the review of strategic planning – the greater metropolitan region.* Obtained from UTS library.

Dobinson, K. (Transport Planner and Advisor, SOCOG). (2008) *Interview:* 'Planning Sydney's transport for sustainability and the 2000 Olympics', Sydney: 15 February 2008.

Dunn, S. (External relations, transportation planning division of ARC). (2009) *Interview:* 'Atlanta's long-term benefits and Olympic transport planning approach', Atlanta, GA: 17 February 2009.

Ericsson, G., Adefope, H., Carrion, R., Elizade, F., Ajan, T., Larfaoui, M., et al. (ca. 1993) *Report IOC Commission for the Games of the XXVII Olympiad 2000.* Obtained from Australian Centre for Olympic Studies (ACOS) at the University of Technology Sydney (UTS). Olympic and event studies collection.★★

Ericsson, G., Adefope, H., Tallberg, P. and Schmitt, P. (1990) *Report of the IOC study and evaluation commission for the preparation of the Games of the XXVIth Olympiad – 1996 on its visit to Atlanta.* Obtained from Olympic Museum (Lausanne, Switzerland). Olympic Studies Centre research archives: CIO / Summer Olympic Games of Atlanta 1996 – transport.★

Frantzeskakis, J. (Transport Planner and Advisor, ATHOC). (2007) *Interview:* 'Transport planning for the Olympic Games in 2004', Athens: 30 August 2007.

Gonzalez, M., Giovanola M., Bovy, P. and Roth, P. (2000) 'de Barcelone 1992 a Athens 2004 – evolution de la structure et des transports des Jeux Olympiques d'ete', unpublished thesis, Ecole Polytechique Federale de Lausanne.

Grant, J. (Manager Transport Planning for the Department of Transportation, NSW). (2008) *Interview:* 'Hombush's strategy and the influence of the IOC', Sydney: 27 February 2008.

Greenpeace. (1995) *Sydney 2000 Transport Action Paper.* Obtained from Australian Centre for Olympic Studies (ACOS) at the University of Technology Sydney (UTS). Olympic and event studies collection.★★

Holsa. (1992a) *Anella Olimpica de Montjuic.* Obtained from Centre for Olympic Studies (CEO) at the Universitat Autònoma de Barcelona (Barcelona, Spain). Research archives: CEO / transport general.★★

— (1992b) *Barcelona '92.* Obtained from Centre for Olympic Studies (CEO) at the Universitat Autònoma de Barcelona (Barcelona, Spain). Research archives: CEO / transport general.★★

IOC (International Olympic Committee). (1999) *Olympic Movement AGENDA 21. Sport for sustainable development.* Obtained from Australian Centre for Olympic Studies (ACOS) at the University of Technology Sydney (UTS). Olympic and event studies collection.★★

— (2003a) *Global Review of the Metro Project.* Obtained from IOC Archives / Summer Olympic Games of Athens 2004 – transport.★

— (2003b) *Global Review of the Suburban Rail Project.* Obtained from IOC Archives / Summer Olympic Games of Athens 2004 – transport.★

— (2003c) *Global Review of the Tram Project.* Obtained from IOC Archives / Summer Olympic Games of Athens 2004 – transport.★

Kalapoutis, A. (Transport Manager, OASA). (2007) *Interview:* 'Transport planning for the 2004 Olympic Games', Athens: 3 September 2007.

Konstantinos, M. (Chief Executive Officer of Athens publit transport authority, OASA). (2007) *Interview:* 'OASA involvement in Olympic Transport Operations', Athens: 3 September 2007.

Macey, T. (2003) *Meltdown subject: event services & venue operations.* Obtained from Audiovisual Archive of CIO / Summer Olympic Games Atlanta 1996 – interview Tim Macey; tape 2 out of 4.★★

MARTA (Metropolitan Atlanta Rapid Transit Authority). (1996) *Handout. Tips for Atlantans riding MARTA during the Games.* Obtained from Subject File Box 2, Kenan Research Centre, Atlanta History Centre Archives.★★

McIntyre, S. (Director of Housing and Community Development, Department of Housing NSW, advisor SOCOG). (2008) *Interview:* 'ORTA's transport planning approach and the influence of the IOC', Sydney: 29 February 2008.

McNevin, N. (Head of town planning within the Olympic Delivery Authority). (2007) *Interview:* 'London's urban development and transport planning approach', London: 11 December 2007.

Millet, L. (MBAassociates, lead architect and advisor, BOCOG). (2007) *Interview:* 'Barcelona's four Olympic areas', Barcelona: 14 September 2007.

Myer, A. (1996) *Millennium park: legacy of the Sydney Olympics.* Green Games Watch 2000. Obtained from the Australian Centre for Olympic Studies (ACOS) at the University of Technology Sydney (UTS). Olympic and event studies collection.★★

OCA (Olympic Coordination Authority). (1995b) *Homebush Bay Rail Link: Statement of Environmental Effects.* Obtained from Australian Centre for Olympic Studies (ACOS) at the University of Technology Sydney (UTS). Olympic and event studies collection.★★

— (1995c) *Homebush Bay Roads Infrastructure: Development Application and Statement of Environmental effects.* Obtained from Australian Centre for Olympic Studies (ACOS) at the University of Technology Sydney (UTS). Olympic and event studies collection.★★

— (1997) *State of play. A report on Sydney 2000 Olympics. Planning & Construction.* Obtained from Australian Centre for Olympic Studies (ACOS) at the University of Technology Sydney (UTS). Olympic and event studies collection.★★

ORTA (Olympic Roads and Transit Authority). (1998a) *98 Annual Report. Moving towards 2000.* Obtained from Australian Centre for Olympic Studies (ACOS) at the University of Technology Sydney (UTS). Olympic and event studies collection.★★

— (1998b) *Sydney 2000 Olympic and Paralympic Games transport strategic plan.* Obtained from Australian Centre for Olympic Studies (ACOS) at the University of Technology Sydney (UTS). Olympic and event studies collection.★★

— (1999a) *Olympic Transport Plan. Transport Operations for the Sydney 2000 Olympic Games.* Obtained from Olympic Museum (Lausanne, Switzerland). Olympic Studies Centre research archives: CIO / Summer Olympic Games of Sydney 2000 – transport.★

— (1999b) *Transport Strategy for the Sydney 2000 Olympic Games.* Obtained from Australian Centre for Olympic Studies (ACOS) at the University of Technology Sydney (UTS). Olympic and event studies collection.★★

Pàmies Jaume, O. (Communications officer for Transports Metropolitan de Barcelona). (2007) *Interview:* 'Public transport development', Barcelona: 13 September 2007.

Papadimitriou, F. (2004) *Transport for the 2004 Olympics.* Obtained from H. Babis.★★

Patrikalakis, K. (Head of public transportation, OASA, ATHOC member). (2007) *Interview:* 'Olympic Operations by OASA', Athens: 3 September 2007.

Protopsaltis, P. (Planning S.A., Head of Transport Division, ATHOC). (2007) *Interview:* 'Transport management operations during the Olympic Games', Athens: 4 September 2007.

Richmond, D. 'Games benefits and legacy – Sustainable Development as it relates to the Games', paper presented at 2012 Applicant City Seminar, Lausanne: 9 October 2003.

Sala-Schnorkowski, M. (President of the Economic and Social Council of Catalunya former president of TMB). (2007) *Interview:* 'Public transport planning for the Olympics', Barcelona: 12 September 2007.

Samaranch, J. A. (1989) *Memoria explicativa de la linea 2 de metro.* Obtained from Research archives: CEO / Olympic Archives / Summer Olympic Games Barcelona 1992 – 'lieu competition to transport position du CIO'.★

Searle, G. (Professor of Urban Planning at the University of Technolgy Sydney). (2008) *Interview:* 'Planning for Sydney and the Olympic Games', Sydney: 13 February 2008.

Sermpis, D. V. (Traffic manager working at TMC Athens during 2004 Olympics). (2007) *Interview:* 'Traffic Management Center Operation and Legacy', Athens: 31 August 2007.

Short, W. (Head of transport for London's organizing committee). (2007) *Interview:* 'London's transport investment and operational system approach', London: 11 December 2007.

SOCOG (Sydney Organizing Committee for the Olympic Games). (1994) *General Oganisation Master Plan.* Obtained from Australian Centre for Olympic Studies (ACOS) at the University of Technology Sydney (UTS). Olympic and event studies collection.★★

TMB (Transports Metropolitans de Barcelona). (1988) *La Linea 2 de Metro – memoria explicativa.* Obtained from Olympic Museum (Lausanne, Switzerland). Olympic Studies Centre research archives: CIO / Summer Olympic Games of Barcelona 1992 – transport.★

— (1989) *Condiciones especial 1989.* Obtained from Olympic Museum (Lausanne, Switzerland). Olympic Studies Centre research archives: CIO / Summer Olympic Games of Barcelona 1992 – transport.★

— (1990) *Accesso del transporte internal.* Obtained from Olympic Museum (Lausanne, Switzerland). Olympic Studies Centre research archives: CIO / Summer Olympic Games of Barcelona 1992 – transport.★

— (1992) *El servicio de TMB y los Juegos Olimpicos 1992.* Obtained from Oriol Pàmies Jaume.★★

Traffic and Transport Directorate. (2001) *DRAFT: Review of the Traffic Management for the 2000 Olympic Games,* Sydney: Roads and Traffic Authority. Obtained from Prof. P. Bovy.★★

Villalante, M. (Director General de Transport Terrestre. Generalitat de Catalunya, Head of Transport Operations, BOCOG). (2007) *Interview:* 'Interview on public transport operations in Barcelona during the 1992 Olympics', Barcelona: 13 September 2007.

Waters, M. (State Traffic Engineer Georgia). (2009) *Interview:* 'Transportation legacy of the Atlanta Olympics', Atlanta, GA: 17 February 2009.

INDEX